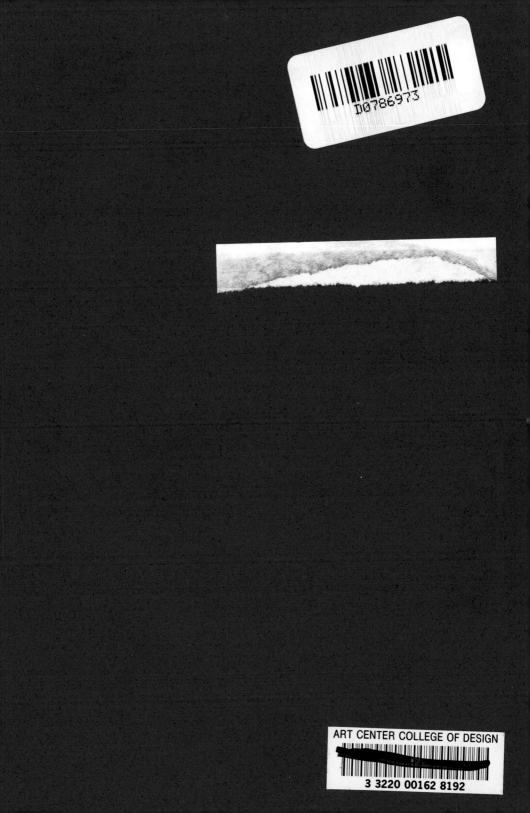

VISUALIZING
BIOLOGICAL
INFORMATION

ART CENTER COLLEGE OF DESIGN

VISUALIZING

BIOLOGICAL

INFORMATION

Editor

Clifford A Pickover

IBM Thomas J. Watson Research Center

World Scientific
Singapore • New Jersey • London • Hong Kong

Published by

World Scientific Publishing Co. Pte. Ltd.

P O Box 128, Farrer Road, Singapore 912805

USA office: Suite 1B, 1060 Main Street, River Edge, NJ 07661

UK office: 57 Shelton Street, Covent Garden, London WC2H 9HE

British Library Cataloguing-in-Publication Data
A catalogue record for this book is available from the British Library.

Library of Congress Cataloging-in-Publication Data
Visualizing biological information / editor, Clifford A. Pickover.
     p.   cm.
    Includes bibliographical references and index.
    ISBN 9810214278 : $64.00
    1. Molecular biology.   2. Molecular genetics.   3. Computer
graphics.   4. Visualization.   I. Pickover, Clifford A.
QH506.V54   1995
574.87'328'011366--dc20   95-47116
                                         CIP

Printed in Singapore by Uto-Print

Preface

Information-Containing Sequences in Biology

"Some people can read a musical score and in their minds hear the music ... Others can see, in their mind's eye, great beauty and structure in certain mathematical functions ... Lesser folk, like me, need to hear music played and see numbers rendered to appreciate their structures."

Peter B. Schroeder
BYTE (1986)

Among the methods available for characterizing information-containing sequences in biology, computer graphics is emerging as an important tool. Having long been applied to the representation of 3-D molecular models, one new area of great potential for computer graphics is the characterization of protein and DNA sequences. DNA (deoxyribonucleic acid) contains the basic genetic information in human and other organisms. Some of this genetic information codes for proteins represented as a sequence of amino acids. Proteins have many different biological functions. For example, they act as enzymes to catalyze chemical reactions, bind molecules for transport, act in contractile and motile systems, and serve as structural elements. The focus of this preface is on DNA, not proteins, because I feel that visual representations of genetic sequences will have greater impact on science in the next few years. However, many fascinating and important chapters discuss graphical methods for displaying information contained in the amino acid sequences of proteins.

Although DNA was first discovered in cell nuclei as long ago as 1869 by F. Miescher, it was not until 1944 that O. Avery and his colleagues identified it as being an information-containing molecule. Interestingly, the amount of DNA per cell is frequently not in proportion to the complexity of the organism, particularly in cells with nuclei. On average, higher plant cells, such as those in the Tuliptree leaf in the frontispiece of this preface (*Liriodendron tulipifera L.*), contain about 2.5 pg per cell, compared with 6 pg for mammals, 7 pg for amphibia, 2 pg for fishes, 1.2 pg for mollusks, and 0.01 pg for bacteria (1 pg = 1×10^{-12} g). There is a wide variation in the amount of repeated DNA in plants. Maize, for example, has an impressive 7.5 pg of DNA in each of its haploid cells. Salamanders have the highest known DNA content for vertebrates (83 pg per cell).

Today, we know that human DNA contains a fantastically long message — equivalent in length to 13 sets of the *Encyclopedia Brittanica*. Written in a language with only four letters, A, C, G, and T — the symbols for the four chemical bases adenine, cytosine, guanine, and thymine — it programs the birth, development, growth, and death of all human inhabitants of Earth. Some stretches define the structure of proteins, others contain a variety of regulatory signals, but the purpose of most of our DNA is unknown. In fact, the protein-coding regions of DNA can be likened to small ships sailing on some vast planetary ocean. We know something about the sailing vessels, but of the waters we know little.

The order of chemical bases in genetic sequences determines the information content of a particular gene or of a piece of DNA. If the human genome were published in a book, its three billion letters would take a third of a lifetime to read. Its sequence of letters would stretch for 5,000 miles if typed out in a single line. Even the much simpler bacterium, *Escherichia coli*, which has a mere 4.8 million bases, would have a sequence of letters which stretches for seven miles. In May 1992, for the first time, molecular biologists were able to peruse the complete DNA sequence of an entire eucaryote chromosome: The 315,357 bases of chromosome III of the yeast, *Saccharomyces cervisiae*. Since then, the number of long sequences has exploded. As of July 1995, the largest contiguous DNA sequence is 2,388 kilobases from the worm *C. elegans.* For other members of the "100 Kilobase Club", a listing that includes DNA sequences greater than 100 Kilobases in length, visit the world wide web site: http://golgi.harvard.edu/100kb/.

Interestingly, the DNA and gene sequences from widely disparate organisms are strikingly similar. For example, more than 400 genes have been identified in the fruit fly that are highly similar to their human counterparts. As another example, the homeobox, part of the control mechanism for development, was initially discovered in the fruit fly, but similar genes have been subsequently discovered in many other organisms. This means that by studying the genetic sequence of any organism we learn something about all organisms. Computerized data bases and computer graphics are now vital in determining the potential significance of, and recognizing patterns in, the plethora of sequences.

In addition to containing regulatory codes, tumor-promoting codes, and codes for protein synthesis, DNA base sequence and composition is often correlated with physical properties of the DNA. For example, the melting temperature of DNA is related to the mole fraction of triple-bonded G-C in the DNA, and the melting transition of synthetic DNAs with regularly alternating sequences is quite sharp. An interesting and common feature of eucaryotic DNA is the presence of tandem as well as interspersed base sequence repeats throughout the genome. These repeating units range in size from dinucleotide repeats to longer interspersed sequences, for example, large sequences known as "Alu sequences" found in higher organisms. Finally, processes of DNA rearrangement, recombination, and a variety of topological changes are all affected by the specific sequence of bases in DNA.

The Human Genome Project

> *"Today literally no biochemical problem can be studied in isolation from its genetic background."*

Albert Lehninger
Biochemistry (1976)

Around 1990, a colossal scientific project known as the human genome project was officially launched with the ambitious goal of understanding the genetic essence of the human animal. The project will take about 15 years to obtain most of its results and costs about US\$3 billion.[1] Our genome consists of 50,000 to 100,000 genes, each gene containing 2,000 to 2 million bases. The human genome project initiative is aimed at determining the location of these genes and analyzing the structure of DNA. More than 50,000 of these genes have been partially deciphered so far.

Already the flood of new data pouring out of laboratories is swamping researchers. The human genome project, along with recent advances in understanding viral sequences that induce cancer and the mechanisms by which oncogenes are activated in tumors, and recent improved gene mapping techniques (with accompanying proliferation of sequence data and increasing commercial interest), will surely make "scientific visualization" of the data a necessity.

Scientific Visualization

> *"Visualization is a method of computing. It transforms the symbolic into the geometric, enabling researchers to observe their simulations and computations. Visualization offers a method for seeing the unseen. It enriches the process of scientific discovery and fosters profound and unexpected insights. In many fields it is already revolutionizing the way scientists do science."*
>
> B. McCormick, T. DeFanti, and M. Brown
> *ACM SIGGRAPH Computer Graphics* (1987)

Computer graphics has become indispensable in countless areas of human activity — from colorful and light-hearted television commercials, to strange new artworks, to evolutionary biology, to processed images from the edges of the known universe. As additional background, the term *scientific visualization* has come to mean the marriage of high-speed computation and colorful 3-D graphics to help researchers better understand complicated models. *Scientific visualization is the art of making the unseen visible.*

A recent National Science Foundation study found that the sciences were in urgent need of government support for graphic tools to view the millions of bytes of data that computers are heaping upon researchers. The commercial world is beginning to recognize the visualization needs of the scientific

---

[1]This figure compares with other proposed "big science" projects: The Hubble Space Telescope (US\$1.5 billion), the Superconducting Supercollider (US\$8 billion), and the Freedom Space Station (US\$30 billion).

community and respond to them. Graphics workstations were born nearly a decade ago to give investigators more comprehensible representations of their results. Manufacturers such as Silicon Graphics, IBM, Evans and Sutherland, and many others have aimed to bridge the gap between fast computation and 3-D images.

Friedhoff and Kiely[2] have noted that the standard argument to promote scientific visualization is that today's researchers must consume ever larger volumes of numbers that gush out, as if from a fire hose, of supercomputer simulations or high-powered scientific instruments. If researchers try to read the genetic sequence data in the form of letters, they will take in the information at a snail's pace. If the information is rendered graphically, however, they can assimilate it at a much faster rate. Therefore, the emphasis of this book is on the graphic representation of information-containing sequences, such as DNA and amino acid sequences, in order to help the human analyst find interesting patterns.

Objectives

My goal is to make this voyage through molecular biology, genetics, and computer graphics as accessible to a broad audience as possible, with the inclusion of glossaries at the end of most chapters, and program outlines where applicable. The book will be of most interest to biologists and computer scientists, and the various reading lists should be of interest to beginners and advanced students of molecular genetics and computer science. Some past methods are reviewed, and interesting avenues for future research in DNA and protein sequence representation are suggested.

Fairly detailed comparisons between DNA and protein sequences are useful and can be achieved by a variety of brute-force statistical computations, but sometimes at the cost of losing an intuitive feeling for the structures. Differences between sequences may obscure the similarities. Even determining whether a particular sequence is *random* is curiously difficult. The best that can be done is to specify certain tests for types of randomness and then to call a sequence random to the degree that it passes them. For example, for DNA one can insist that each base occurs with frequency 1/4. Of course, this does not test for the spatial progression of the bases — and permutations of

---

[2]Friedhoff, R. and Kiely, T., "The eye of the beholder", *Computer Graphics World* **13**(8), 47–56 (1990).

bases taken two at a time, three at a time, ... , n at a time must also be checked. The importance of "randomness" in studying sequence data (and in understanding the implications for evolution) is discussed in many papers.[3] These difficulties also motivate the use of computer graphics as an aid to statistical computations.

In the chapters which follow, researchers have used both 2-D and 3-D graphics to help visualize a range of phenomena. In general, they have sought to find pattern and meaning in the cacophony of genetic and protein sequence data using unusual computer graphics and audio techniques. The usefulness of a particular graphic representation and audio technique is determined by its descriptive capacity, potential for comparison, aid in focusing attention, and versatility. For background articles in the field of scientific visualization, see some of the references in the following sections. The figures in this preface briefly indicate several graphical representations which I have experimented with over the last few years.

Clifford A. Pickover
Yorktown Heights, New York
1995

---

[3] "Eukaryotes, prokaryotes: Who's first?" *Science News* **129**, 280 (1986); Lewin, R., Computer genome is full of junk DNA", *Science* **232**, 577–578 (1986).

General Reading

Computers, Genetics, Proteins

These books and papers are for a general audience and all address the use of computers in understanding genetic sequences. Many articles touch on the human genome project.

Angier, N., "Biologists seek the words in DNA's unbroken text", *New York Times*, Tuesday, July 9, p. C1 (1991).[4]

Beugelsdijk, T., "Engineering the human genome project", *IEEE Potentials* **9**(4), 34–38 (1990).

Courteau, J., "Genome databases", *Science* **254**(5029), 201–207 (1991).

Erickson, D., "Hacking the genome", *Scientific American* **266**(4), 128 (1992).[5]

Frenkel, K., "The human genome project and informatics", *Communications of the ACM* **34**(11), 40 (1991).

Friedland, P. and Kedes, L., "Discovering the secrets of DNA", *Communications of the ACM* **28**, 1164–1186 (1985).

Lander, E., Langridge, R. and Saccocio, D., "A report on computing in molecular biology: Mapping and interpreting biological information", *Communications of the ACM* **34**(11), 32 (1991).

Lowenstein, J., "Whose genome is it, anyway?" *Discover* **13**(5), 28–30 (1992).

Manzel, J., "Supercomputing in biomedical research: Applications to sequence analysis and molecular biology", *Cray Channels* Fall, 2–5 (1988).

---

[4]Linguistic methods help experts pinpoint key genetic instructions amid the biochemical babble.

[5]The masses of information written in human DNA are being translated into the precise ones and zeros of computer code. The choices made now will determine whether the genome project is useful for future discovery or just a binary junkyard of data.

Pearson, P., Maidak, B., Chipperfield, M. and Robbins, R., "The human genome initiative — Do databases reflect current progress?" *Science* **254**(5029), 214 (1991).

The Human Genome Project.[6]

General Computational Biology

Branden, C. and Tooze, J., *Introduction to Protein Structure* (Garland, 1991).

Cantor, C. and Schimmel, P., *Biophysical Chemistry* (W. H. Freeman and Company, 1980).

Cinkosky, J., Fickett, J., Gilna, P. and Burks, C., "Electronic data publishing and GenBank", *Science* **252** (May 31), 1273–1277 (1991).

Daintith, J., *A Concise Dictionary of Chemistry* (Oxford University Press, 1990).

Dickerson, R. and Geis, I., *The Structure and Action of Proteins* (Benjamin Cummings, 1989).

DeLisi, C., "Computers in molecular biology: Current applications and emerging trends", *Science* **24** (April 1), 47–52 (1992).

Human Genome News.[7]

Lehninger, A., *Biochemistry* (Worth, 1976).

Lipton, R., Marr, T. and Welsh, J., "Computational approaches to discovering semantics in molecular biology", *Proceedings of the IEEE* **77**(7), 1056–1060 (1989).

---

[6]A brochure is available from National Center for Human Genome Research, Building 38A, Room 617, 9000 Rockville Pike, Bethesda, MD 20892, USA.

[7]A free newsletter published by the Department of Energy every other month. Contact: Betty K. Mansfield, Oak Ridge National Laboratory, 1060 Commerce Park, MS 6480, Oak Ridge, TN 37830, USA. Also see the web site: http://golgi.harvard.edu/htbin/biopages?genome+news.

Merriam, J., Ashburner, M., Hartl, D. and Kafatos, F., "Towards cloning and mapping the genome of Drosphilia", *Science* **254**(5029), 221–225 (1991).[8]

Human Genome 1989-90 Program Report and *Understanding Our Genetic Inheritance, The U.S. Human Genome Project: The First Five Years FY 1991–1995.*[9]

Roskies, R., "Supercomputing and biomedical science", *Future Generations Computer Systems* (September 5), 197–205 (1989).

Shalloway, D., "Information management in recombinant DNA research", *Perspectives in Computing* **5**(3/4), 22–33 (1985).

Voet, D. and Voet, J., *Biochemistry* (Wiley, 1990).

Walker, P., *Cambridge Dictionary of Biology* (Cambridge University Press, 1989).

Wasserman, S. and Cozzarelli, N., "Biochemical topology: Application to DNA recombination and replication", *Science* **232**, 951–955 (1986).

Watson, J., Hopkins, N., Roberts, J., Steitz, J. and Weiner, A., *Molecular Biology of the Gene* (Benjamin Cummings, 1987).

Using Computer Graphics to Represent DNA and Protein Sequences

Listed here are unusual, useful references concerning the graphic representation of DNA and protein sequences in order to find biologically relevant patterns. For a general background, see issues in the *Journal of Molecular Graphics* (Butterworth and Company).

---

[8]The size of the *D. Melanogaster* genome is estimated to be 165 Mb compared with 3000 Mb for the human genome. Estimates for the number of genes range from about 5,000 to more than 15,000.

[9]Oak Ridge reports. Contact: Betty K. Mansfield, Oak Ridge National Laboratory, P.O. Box 2008, Oak Ridge, TN 37831-6050, USA.

Burks, C., "Towards modeling DNA sequences as automata", *Physica* **10D**, 157–167 (1984).

Churchill, G., "Stochastic models for heterogeneous DNA sequences", *Bulletin of Mathematical Biology* **41**(1), 79–94 (1989).

Collins, J. and Coulson, A., "Application of parallel processing algorithms for DNA sequence analysis", *Nucleic Acids Research* **12**, 181–192 (1984).

Cowan, J., Jellis, C. and Rickwood, D., "A new method of representing DNA sequences which combines ease of visual analysis with machine readability", *Nucleic Acids Research* **14**(1), 509–520 (1986).

Gates, M., "A simple way to look at DNA", *Journal of Theoretical Biology* **119**, 319–328 (1986).

Gibbs, A. and McIntyre, G., "The diagram: A method for comparing sequences. Its use with amino acid and nucleotide sequences", *European Journal of Biochemistry* **16**, 1–11 (1970).

Harr, R., Hagblom, P. and Gustafsson, P., "Two-dimensional graphic analysis of DNA sequence homologies", *Nucleic Acids Research* **10**, 365–374 (1982).

Hamori, E. and Ruskin, J., "H-curves, a novel method of representation of nucleotide series especially suited for long DNA sequences", *Journal of Biological Chemistry* **258**(2), 1318–1327 (1983).

Hamori, E., "Novel DNA sequence representations", *Nature* **314**, 585–586 (1985).

Hamori, E., "Graphics representation of long DNA-sequences by the method of H-curves: Current results and future aspects", *Biotechniques* **7**(7), 710–720 (1989).

Hamori, E., Varga, G. and LaGuardia, J., "HYLAS: Program for generating H-curves (abstract three-dimensional representations of long DNA sequences)", *Computer Applications in Biological Science* **5**(4), 263–269 (1981).

Hamori, E., "Visualization of biological information encoded in DNA", *Frontiers of Scientific Visualization*, eds. C. Pickover and S. Tewksbury (Wiley, 1994).

Jeffrey, H., "Chaos game representation of gene structure", *Nucleic Acids Research* **18**(8), 2163–2170 (1990).

Jeffrey, H., "Chaos game representation of sequences", *Computers and Graphics* **16**(1), 25–33 (1992).

Jeffrey, H., "Fractals and genetics of the future", *Visions of the Future*, ed. C. Pickover (St. Martin's Press, 1993).

Lathe, R. and Findlay, R., "Novel DNA sequence representations", *Nature* **314**, 585–586 (1985).[10]

Lewin, R., "Proposal to sequence the human genome stirs debate", *Science* **232**, 1598–1599 (1986).

Li, W. and Kunihiko, K., "Long-range correlation and partial $1/f^\alpha$ spectrum in a non-coding DNA sequence", *Santa Fe Institute Report*, July 30 (1991).

Melcher, U., "A readable and space-efficient DNA sequence representation: Application to caulimoviral DNAs", *Computer Applications in the Biosciences* **4**, 93–96 (1988).

Mizraji, E. and Ninio, J., "Graphical coding of nucleic acid sequences", *Biochimie* **67**, 445–448 (1985).

Ninio, J., "The sequence explosion: The crazy years 1980–1990", *Biochemical Systematics and Ecology* **11**(3), 305–313 (1983). (In French)

Novotny, J., "Matrix program to analyze primary structure homology", *Nucleic Acids Research* **10**, 127–131 (1982).

Peng, C-K., *et al.*, "Long-range correlations in nucleotide sequences", *Nature* **356**(6356), 168–170 (1992).[11]

---

[10] A reply to Hamori.

[11] With the help of 2-D representations of DNA sequence and statistical analysis, the authors show long-range correlations in genomic sequences but not in cDNA.

Pickover, C., "DNA and protein tetragrams: Biological sequences as tetrahedral movements", *Journal of Molecular Graphics* **10**(1), 2–6 (1992).

Pickover, C., *Computers, Pattern, Chaos and Beauty* (St. Martin's Press, 1990). (Chapter on DNA vectorgrams)

Pickover, C., "DNA vectorgrams: Representation of cancer gene sequences as movements along a 2-D cellular lattice", *IBM Journal of Research and Development* **31**, 111–119 (1987).

Pickover, C., "Computer-drawn faces characterizing nucleic acid sequences", *Journal of Molecular Graphics* **2**, 107–110 (1984).

Pickover, C., "Frequency representations of DNA sequences: Application to a bladder cancer gene", *Journal of Molecular Graphics* **2**, 50 (1984).

Pickover, C., "On genes and graphics", *Speculations in Science and Technology* **12**(1), 5–15 (1990).

Schneider, K., "Gene mapping is improved", *New York Times*, June 26, Sec. D2 (1990).

Silverman, B.D. and Linsker, R., "A measure of DNA periodicity", *Journal of Theoretical Biology* **118**, 295–300 (1986).

Swanson, R., "A vector representation for amino acid sequences", *Bulletin of Mathematical Biology* **46**(4), 623–639 (1984).

Swanson, R., "A unifying concept for the amino acid code", *Bulletin of Mathematical Biology* **46**, 187–203 (1984).

Zhang, C. and Zhang, R., "Diagrammatic representation of the distribution of DNA bases and its applications", *International Journal of Biological Macromolecules* **13**, 45–49 (1991).

Genetics and Music

"Organisms which have evolved on this earth are governed by multitudes of periodicities. Individual genes have been duplicated often to the point of redundancy. This principle even appears to govern the manifestations of human intellect; musical compositions also rely on repetitious recurrence."

Susumu and Midori Ohno
Immunogenetics (1986)

Hayashi, K. and Munakata, N., "Basically musical", *Nature* **310**, 96 (1984).

Ohno, S., "Modern coding sequences are in the periodic-to-chaotic transition", *Modern Trends in Human Leukemia VIII* (Springer, 1989) pp. 512–519.[12]

Ohno, S. and Ohno, M., "The all persuasive principle of repetitious recurrences governs not only coding sequence construction but also human endeavor in musical composition", *Immunogenetics* **24**, 71–78 (1986).[13]

Ohno, S., "Codon preference is but an illusion created by the construction principle of coding sequences", *Proceedings of the National Academy of Science* **85**, 4328–4378 (1988).

Ohno, S., "Of words, genes and music", *The Semiotics of Cellular Communication in the Immune System, NATO ISI Series*, Vol. H32, ed. E. Sercarz (Springer, 1988) pp. 131–147.

Ohno, S., "The grammatical rule of DNA language: Messages in palindromic verses", *Evolution of Life, Fossils, Molecules and Culture*, eds. S. Osawa and T. Honjo (Springer, 1991).

Ohno, S. and Yomo, T., "The grammatical rule for all DNA: Junk and coding sequences", *Electrophoreses* **12**, 103–108 (1991).

Pickover, C., "There is music in our genes", *Mazes for the Mind: Computers and the Unexpected* (St. Martin's Press, 1992).

---

[12]Includes a two-page musical score for the musical transformation of a part of the chicken αA crystalline coding sequence.
[13]Contains four pages of musical scores.

"*.... Both genes and music are made of linear and quantized information which represent unfathomable diversity and mystery. However, we are not confident about how to disentangle the intricate logic of life's composition.*"

Naobuo Munakata and Kenshi Hayashi
Mazes for the Mind (1992)

Genetics and Fractals

"*It's almost incredible that the occupant of one site on a gene would somehow influence which nucleotide shows up 100,000 bases away.*"

H. Eugene Stanley

There has been significant recent work in the fractal structure and long-range correlation characteristics of DNA sequences. For a recent review article, see:

Amato, I., "DNA shows unexplained patterns", *Science* **257**(5071), 747 (1992).[14]

For another review of this work see:

Yam, P., "Noisy nucleotides: DNA sequences show fractal correlations", *Scientific American* **267**(3), 23–27 (1992).

For other work in this area, see:

Jeffrey, H., "Chaos game representation of gene structure", *Nucleic Acids Research* **18**(8), 2163–2170 (1990).

Jeffrey, H., "Chaos game representation of sequences", *Computers and Graphics* **16**(1), 25–33 (1992).

---

[14]The article describes the work of W.-T. Li of Rockefeller University and K. Kaneko of the University of Tokyo who studied long-range patterns in DNA. Also described is the work of IBM physicist Richard Voss who reports self-similarity of DNA sequences after an analysis of 25,000 genes.

Jeffrey, H., "Fractals and genetics of the future", *Visions of the Future*, ed. C. Pickover (St. Martin's Press, 1993).

Li, W. and Kunihiko, K., "Long-range correlation and partial $1/f^\alpha$ spectrum in a non-coding DNA sequence", *Santa Fe Institute Report*, July 30 (1991).

Pickover, C., "DNA and protein tetragrams: Biological sequences as tetrahedral movements", *Journal of Molecular Graphics* **10**(1), 2–6 (1992).

Scientific Visualization

General

> *"Today's data sources are such fire hoses of information that all we can do is gather and warehouse the numbers they generate."*
>
> B. McCormick, T. DeFanti, and M. Brown
> *ACM SIGGRAPH Computer Graphics* (1987)

Foley, J. and van Dam, A., *Fundamentals of Interactive Computer Graphics* (Addison-Wesley, 1982).

Friedhoff, R. and Benzon, W., *Visualization: The Second Computer Revolution*, (Abrams, 1989).

Hargittai, I. and Pickover, C., *Spiral Symmetry* (World Scientific, 1992).[15]

McCormick, B., DeFanti, T. and Brown, M., "Visualization in scientific computing", *ACM SIGGRAPH Computer Graphics* **21**(6), (1987).[16]

Pickover, C., *Computers, Pattern, Chaos, and Beauty* (St. Martin's Press, 1990).

Pickover, C., *Computers and the Imagination* (St. Martin's Press, 1991).

---

[15]Spirals in nature, biology, astronomy, botany, mathematics, and art.

[16]This report is accompanied by two hours of videotape. To request copies: ACM Order Dept., P.O. Box 64145, Baltimore, MD 21264, USA. For other visualization tapes and catalogs, contact: http://www.siggraph.org/library/SVR/SVR.html.

Pickover, C. and Tewksbury, S., *Frontiers of Scientific Visualization* (Wiley, 1994).

Pickover, C., *Mazes for the Mind: Computers and the Unexpected* (St. Martin's Press, 1992).

Tufte, E., *The Visual Display of Quantitative Information* (Graphics Press, 1983).

Wainer, H. and Thissen, D., "Graphical data analysis", *Annual Review of Psychology* **32**, 191–241 (1981).

Wolff, R., "The visualization challenge in the physical sciences", *Computers in Science* **2**(1), 16–31 (1988).

Journals

There are many journals devoted to the subject of scientific visualization including *Computers and Graphics, The Visual Computer, IEEE Computer Graphics and Applications, The Journal of Visualization and Computer Animation, Computer Graphics World,* and *Pixel.*

Federal Research Organizations Involved with Computational Genetics

U.S. Department of Energy, Human Genome Project, Washington, DC 20545, USA.

National Institute of Health, Building 38A, Bethesda MD 20894, USA.

National Science Foundation, Database Programs, Division on Information, Robotics, and Intelligent Systems, 1800 G Street N.W., Washington, DC 20550, USA.

National Library of Medicine, Building 38A, Bethesda, MD 20894, USA.

For information on genome research in Europe, see: Bodmer, W. "Genome research in Europe", *Science* **256**(5056), 480–481 (1992). The United Kingdom, Denmark, France, Germany, and Italy have established national genome programs, while others, for example, in Sweden, are being developed. Some relevant agencies include HUGO (Human Genome Organization), CEPH (Centre d'Etude du Polymorphism Humain), AFM (French Muscular Dystrophy Association), DKFZ (German Cancer Center), EUROGEM (a consortium of 23 laboratories in ten countries), and ESF (European Science Foundation).

Some Genetic and Biological Database Repositories

Two major repositories for human genome information are the Genome Data Base (GDB) at Johns Hopkins University, for map information, and GenBank at the National Library of Medicine, for DNA sequence information. DNA sequence information has also been assembled in associated databases, the European Molecular Biology Data Library (EMBL) and the DNA Data Bank of Japan (DDBJ). The tradition has been established of submitting sequence information to the databases at the time of publication in a scientific journal. Pearson *et al.* (1991) note that, more recently, submission of sequence data to the databases occurs without any formal publication in a journal, and in fact many journals no longer accept papers concerned primarily with the DNA sequence information. To date, these databases contain more than 120 million sequenced DNA and RNA base pairs from 3,000 species (Corteau, 1991). Here is a list of databases.

NCBI-GenBank, National Center for Biotechnology Information, National Library of Medicine, 38A, 8N805, 8600 Rockville Pike, Bethesda, MD 20894, USA. Phone: 301-496-2475, Fax: 301-480-9241, CD-ROM available, E-mail: info@ncbi.nlm.nih.gov.[17]

EMBL Data Library, Postfach 10.2209, Meyerhofstrasse 1, 6900 Heidelberg, Germany. Fax: 49-6221-387-519, Phone: 49-6221-387-258, E-mail: DataLib@EMBL-Heidelberg.DE.

DDJB, Takashi Gojobori, National Institute of Genetics, Mishima, Shizuoka 411, Japan. Fax: 81-559-75-6040, Phone: 81-559-75-0771, E-mail: ddbj@ddbj.nig.ac.jp.

---

[17]By the time this book is published, NCBI should be providing NetEntrez and NetBlast, an essentially free client-server sequence retrieval and database searching service.

GDB/OMIM User support, Welch Medical Library, 18c0 E. Monument St., 3rd Floor, Baltimore, MD 21205, USA. Fax: 301-955-0054, Phone: 301-955-7058, E-mail: help@welch.jhu.edu. (Human gene mapping information and human genetic traits. OMIM stands for "On-Line Mendelian Inheritance in Man".)

GENINFO, NLM Medlars Management Service, 8600 Rockville Pike, Bethesda, MD 20894, USA. Phone: 800-638-8480, E-mail: info@ncbi.nim.nih.gov. (Nucleotide sequences, protein sequences, and MEDLINE bibliographic abstracts.)

NBRF/PIR, Kathryn Sidman, National Biomedical Research Foundation, 3900 Reservoir Rd., NW, Washington, DC 20007, USA. Fax: 202-687-1662, Phone: 202-687-1662, E-mail: pirmail@gunbrf.bitnet. (30,000 amino acid sequences. PIR stands for "Protein Identification Resource".)

MIPS, Martinsried Institute of Protein Sequences, Hans-Werner Mewes, MPI/GEN, Max-Planck-Institut für Biochemie, 8033 Martinsried, Germany. Fax: 49-89-8578-2655, Phone: 49-89-8578-1, E-mail: mewes@dm0mpb51.bitnet. (30,000 amino acid sequences.)

JIPID, Japanese International Protein Information Database, Akira Tsugita, Research Institute for Biosciences, Science University of Tokyo, Yamazaki, Noda 278, Japan. Fax: 81-471-22-1544, Phone: 81-471-23-9777, E-mail: tsugita@jpnsut31.bitnet. (30,000 amino acid sequences.)

PDB, Protein Data Bank, Chemistry Department, Bldg. 555, Brookhaven National Laboratory, Upton, NY 11973, USA. Fax: 516-282-5815, Phone 516-282-3629, E-mail: pdb@bnlchm.bitnet or pdb@pdb.pdb.bnl.gov. (3-D atomic coordinates; over 1000 entries.)

BIOMAGRESBANK, B.R. Seavy or J.L. Markley, Biochemistry Department, University of Wisconsin, Madison, WI 53706-1569, USA. Fax: 608-262-3453, Phone: 608-263-9349. (NMR spectroscopic data on proteins and protein fragments.)

CSD, Olga Kennard, Cambridge Crystallographic Data Centre, University Chemistry Laboratory, Cambridge CB2 1EW, UK. Fax: 44-223-336408, Phone: 44-223-336408, E-mail: dgw1@uk.ac.cam.phx. (Crystal data, 3-D coordinates, references, molecular topologies.)

CCSD and CARBANK, Dana Smith, Complex Carbohydrate Research Center, University of Georgia, Athens, GA 30602, USA. Fax: 404-542-4412, Phone: 404-542-4484, E-mail: CarbBank@UGA.bitnet or 76060.1127@Compuserve.com. (Polysaccharide and glycopeptide structures.)

BIOSIS User Services, 2100 Arch St., Philadelphia, PA 19103-1399, USA. Fax: 215-587-2016, Phone: 800-523-4806. (Abstracts of the biological literature.)

MEDLARS Management Section, NLM, Bldg. 38, Rm. 4N421, 8600 Rockville Pike, Bethesda, MD 20894, USA. Fax: 301-496-0822, Phone: 800 638-8480. (Biomedical literature relevant for human genome research.)

MOUSE, Janice Ormsby, Jackson Laboratory, 600 Main Street, Bar Harbor, ME 04609, USA. Fax: 207-288-5079, Phone: 207-288-3371 (Ext. 1394), E-mail: davidnaman@jax.org. (Encyclopedia of the mouse genome.)

MOUSE: GBASE, Thomas Roderick, Jackson Laboratory, 600 Main Street, Bar Harbor, ME 04609, USA. Fax: 207-288-5079, Phone: 207-288-3371, E-mail: thr@morgan.jax.org. (Mouse genetic map data.)

E. COLI: ECD, EMBL Data Library, Postfach 10.2209 Meyerhofstrasse 1, 6900 Heidelberg, Germany. Fax: 49-6221-387-519, Phone: 49-6221-387-258, E-mail: DataLib@ EMBL-Heidelberg.DE. (E. coli nucleotide sequence data. EMBL Sequence Data Base also available on CD-ROM.)

For other organism data bases, such as WCS (Worm Community System), E. COLI: CGSC, E. COLI: ECOSEQ, ECOMAP, ECOGENE, NEMATODE: "acedb", DROSOPHILA INFORMATION DATA BASE, DROSOPHILA GENETIC MAPS, and PLANTS, see Courteau (1991).

For various databases of databases, such as DBIR (Directory of Biotechnology Information Resources) and LIMB (Listing of Molecular Biology Databases), see Courteau (1991).

Some of these databases provide special graphic features for displaying genetic maps and related information. These include NEMATODE: acedb, CIS, MOUSE: ENCYCLOPEDIA OF THE MOUSE GENOME, MOUSE: GBASE, CCSD, and CARBBANK.

The BIOSCI Electronic Newsgroup Network

The BIOSCI newsgroup network was developed to allow easy worldwide communications between biological scientists who work on a variety of computer networks.[18] By having distribution sites or "nodes" on each major network, BIOSCI allows its users to contact people around the world without having to learn a variety of computer addressing tricks. Any user can simply post a message to a regional BIOSCI node, and copies of that message will be distributed automatically to all other subscribers on all of the participating networks, including Internet, USENET, BITNET, EARN, NETNORTH, HEANET, and JANET.

For general information on biology, genetic sequences, and computers, the following internet bulletin boards may be of interest:

bionews@net.bio.net (for information on meetings and courses);
bioforum@net.bio.net (miscellaneous).

Please send subscription requests to biosci@net.bio.net and not to the newsgroup posting addresses.

---

[18]The BIOSCI information is provided by Dave Kristofferson, BIOSCI/bionet Manager, E-mail: Kristoff@net.bio.net.

List of BIOSCI Newsgroups

Newsgroup Name	Topic
ACEDB-SOFT	Discussions by users and developers of genome databases using the ACEDB software
AGEING	Discussions about ageing research
AGROFORESTRY	Discussions about agroforestry research
ARABIDOPSIS	Newsgroup for the Arabidopsis Genome Project
BIOFORUM	Discussions about biological topics for which there is not yet a dedicated newsgroup
BIOLOGICAL-INFORMATION-THEORY-AND-CHOWDER-SOCIETY	Applications of information theory to biology
BIONAUTS	Question/answer forum for help using electronic networks, locating e-mail addresses, etc.
BIONEWS	General announcements of widespread interest to biologists
BIO-JOURNALS	Tables of Contents of biological journals
BIO-MATRIX	Applications of computers to biological databases
BIO-SOFTWARE	Information on software for the biological sciences
BIOTHERMOKINETICS	Discussions about the kinetics, thermodynamics and control of biological processes at the cellular level
CELL-BIOLOGY	Discussions about cell biology including cancer research at the cellular level
CHLAMYDOMONAS	Discussions about the biology of the green alga Chlamydomonas and related genera
CHROMOSOMES	Discussions about mapping and sequencing of eucaryote chromosomes
COMPUTATIONAL-BIOLOGY	Mathematical and computer applications in biology
DROSOPHILA	Discussions about biological research on Drosophila
EMBL-DATABANK	Messages to and from the EMBL database staff
EMPLOYMENT	Job opportunities in biology
GDB	Messages to and from the Genome Data Bank staff
GENBANK-BB	Messages to and from the GenBank database staff
GENETIC-LINKAGE	Newsgroup for genetic linkage analysis
HIV-MOLECULAR-BIOLOGY	Discussions about the molecular biology of HIV
HUMAN-GENOME-PROGRAM	NIH-sponsored newsgroup on human genome issues
IMMUNOLOGY	Discussions about research in immunology
INFO-GCG	Discussions about the GCG sequence analysis software
JOURNAL-NOTES	Practical advice on dealing with professional journals
METHODS-AND-REAGENTS	Requests for information and lab reagents
MOLECULAR-EVOLUTION	Discussions about research in molecular evolution

NEUROSCIENCE	Discussions about research in the neurosciences
N2-FIXATION	Discussions about biological nitrogen fixation
PHOTOSYNTHESIS	Discussions about photosynthesis research
PLANT-BIOLOGY	Discussions about research in plant biology
POPULATION-BIOLOGY	Discussions about research in population biology
PROTEIN-ANALYSIS	Discussions about research on proteins and messages for the PIR and SWISS-PROT databank staff
PROTEIN-CRYSTALLOGRAPHY	Discussions about crystallography of macromolecules and messages for the PDB staff
RAPD	Discussions about Randomly Amplified Polymorphic DNA
SCIENCE-RESOURCES	Information from/about scientific funding agencies
TROPICAL-BIOLOGY	Discussions about research in tropical biology
VIROLOGY	Discussions about research in virology
WOMEN-IN-BIOLOGY	Discussions about issues concerning women biologists
YEAST	Discussions about the molecular biology and genetics of yeast

Electronic Mail Addresses of Newsgroups

Newsgroup Name	Mailing Address
ACEDB-SOFT	acedb@net.bio.net
AGEING	ageing@net.bio.net
AGROFORESTRY	ag-forst@net.bio.net
ARABIDOPSIS	arab-gen@net.bio.net
BIOFORUM	bioforum@net.bio.net
BIO-INFORMATION-THEORY	bio-info@net.bio.net
BIONAUTS	bio-naut@net.bio.net
BIONEWS	bionews@net.bio.net
BIO-JOURNALS	bio-jrnl@net.bio.net
BIO-MATRIX	biomatrx@net.bio.net
BIO-SOFTWARE	bio-soft@net.bio.net
BIOTHERMOKINETICS	btk-mca@net.bio.net
CELL-BIOLOGY	cellbiol@net.bio.net
CHLAMYDOMONAS	chlamy@net.bio.net
CHROMOSOMES	biochrom@net.bio.net
COMPUTATIONAL-BIOLOGY	comp-bio@net.bio.net
DROSOPHILA	dros@net.bio.net
EMBL-DATABANK	embl-db@net.bio.net

EMPLOYMENT	biojobs@net.bio.net
GDB	gdb@net.bio.net
GENBANK-BB	genbankb@net.bio.net
GENETIC-LINKAGE	gen-link@net.bio.net
HIV-MOLECULAR-BIOLOGY	hiv-biol@net.bio.net
HUMAN-GENOME-PROGRAM	gnome-pr@net.bio.net
IMMUNOLOGY	immuno@net.bio.net
INFO-GCG	info-gcg@net.bio.net
JOURNAL-NOTES	jrnlnote@net.bio.net
METHODS-AND-REAGENTS	methods@net.bio.net
MOLECULAR-EVOLUTION	mol-evol@net.bio.net
NEUROSCIENCE	neur-sci@net.bio.net
N2-FIXATION	n2fix@net.bio.net
PHOTOSYNTHESIS	photosyn@net.bio.net
PLANT-BIOLOGY	plantbio@net.bio.net
POPULATION-BIOLOGY	pop-bio@net.bio.net
PROTEIN-ANALYSIS	proteins@net.bio.net
PROTEIN-CRYSTALLOGRAPHY	xtal-log@net.bio.net
RAPD	rapd@net.bio.net
SCIENCE-RESOURCES	sci-res@net.bio.net
TROPICAL-BIOLOGY	trop-bio@net.bio.net
VIROLOGY	virology@net.bio.net
WOMEN-IN-BIOLOGY	womenbio@net.bio.net
YEAST	yeast@net.bio.net

Electronic Mail Addresses of USENET Newsgroups

Newsgroup Name	Usenet Newsgroup Name
ACEDB-SOFT	bionet.software.acedb
AGEING	bionet.molbio.ageing
AGROFORESTRY	bionet.agroforestry
ARABIDOPSIS	bionet.genome.arabidopsis
BIOFORUM	bionet.general
BIO-INFORMATION-THEORY	bionet.info-theory
BIONAUTS	bionet.users.addresses
BIONEWS	bionet.announce
BIO-JOURNALS	bionet.journals.contents
BIO-MATRIX	bionet.molbio.bio-matrix
BIO-SOFTWARE	bionet.software

BIOTHERMOKINETICS	bionet.metabolic-reg
CELL-BIOLOGY	bionet.cellbiol
CHLAMYDOMONAS	bionet.chlamydomonas
CHROMOSOMES	bionet.genome.chromosomes
COMPUTATIONAL-BIOLOGY	bionet.biology.computational
DROSOPHILA	bionet.drosophila
EMBL-DATABANK	bionet.molbio.embldatabank
EMPLOYMENT	bionet.jobs
GDB	bionet.molbio.gdb
GENBANK-BB	bionet.molbio.genbank
GENETIC-LINKAGE	bionet.molbio.gene-linkage
HIV-MOLECULAR-BIOLOGY	bionet.molbio.hiv
HUMAN-GENOME-PROGRAM	bionet.molbio.genome-program
IMMUNOLOGY	bionet.immunology
INFO-GCG	bionet.software.gcg
JOURNAL-NOTES	bionet.journals.note
METHODS-AND-REAGENTS	bionet.molbio.methds-reagnts
MOLECULAR-EVOLUTION	bionet.molbio.evolution
NEUROSCIENCE	bionet.neuroscience
N2-FIXATION	bionet.biology.n2-fixation
PHOTOSYNTHESIS	bionet.photosynthesis
PLANT-BIOLOGY	bionet.plants
POPULATION-BIOLOGY	bionet.population-bio
PROTEIN-ANALYSIS	bionet.molbio.proteins
PROTEIN-CRYSTALLOGRAPHY	bionet.xtallography
RAPD	bionet.molbio.rapd
SCIENCE-RESOURCES	bionet.sci-resources
TROPICAL-BIOLOGY	bionet.biology.tropical
VIROLOGY	bionet.virology
WOMEN-IN-BIOLOGY	bionet.women-in-bio
YEAST	bionet.molbio.yeast

BIOSCI Prototype Newsgroups

Posting Address	Purpose
pep-libs@net.bio.net	Discussions on the generation and use of peptide molecular repertoires displayed on phage or prepared as synthetic peptide combinatorial libraries
rna@net.bio.net	Discussions about RNA editing, RNA splicing, and ribozyme activities of RNA
yac@net.bio.net	Dicussions about yeast artificial chromosomes

As with all BIOSCI matters, please address further questions to biosci@net.bio.net.

Sequence Analysis Software

There has been a recent explosion in the number of sequence analysis software products on the market in recent years. For example, the GenBank Software Clearinghouse (a database of software that used the GenBank database) once listed over a hundred packages that perform a range of analyses. Listed in the following are several examples to give a flavor of the range of functions provided by these tools.[19]

1. *CLONES* (database: store, catalog, and access recombinant clones/libraries), (IBM PC, MS-DOS/PC-DOS IBM PS/2, MS-DOS/PC-DOS), Hal B. Jenson, Department of Pediatrics, University of Texas Health Science Center, 7703 Floyd Curl Dr., San Antonio, TX 78284-7811, USA.

2. *Delila System (DEoxyribonucleic acid LIbrary LAnguage)*, (Information Theory Analysis of binding sites), (any computer with Pascal or C compilers), Tom Schneider, Laboratory of Mathematical Biology, National Cancer Institute, FCRDC Bldg. 469, Rm. 144, P.O. Box B, Frederick, MD 21702-1201, USA.

3. *DIGISEQ, TYPESEQ* (entry of DNA sequence data by digitizer or keyboard), (IBM PC, MS-DOS/PC-DOS, IBM PS/2, MS-DOS/PC-DOS), Hal B. Jenson, Department of Pediatrics, University of Texas Health Science Center, 7703 Floyd Curl Dr., San Antonio, TX 78284-7811, USA.

4. *DM5* (sequence analysis software), (IBM PC, MS-DOS/PC-DOS IBM PS/2, MS-DOS/PC-DOS, Sun-3, Unix), kmwillia@arizrvax (bitnet), Genetics Software Center, Molecular and Cellular Biology, University of Arizona, Biosciences West 230, Tucson, AZ 85721, USA.

5. *DNA Parrot Talking Gel Reader* (analyzes gels to find the base sequence), (IBM PC, MS-DOS/PC-DOS, Macintosh), Thomas Maler, T&T Research, 44 George St., Etobicoke, ON M8V 2S2, Canada.

6. *DNAMAT* (IBM Mainframe, VM/CMS), Ron Unger, Computer Science, Weizman Institute of Science, Rehovot 76100, Israel.

7. *FASTA* (rapid sequence comparison), (all computers and operating systems), William R. Pearson, Department of Biochemistry, University of Virginia, Box 440 Jordan Hall, Charlottesville, VA 22908, USA.

---

[19]The software tools in this section are listed for illustrative purposes only. *The editor does not endorse any particular software, company, or product*, nor does he accept responsibility for the selection of any products by the reader. Selections were made from a list compiled by the now defunct GenBank Software Clearinghouse. As this book goes to press, more and more massively parallel programs are emerging for established parallel hardware (e.g., Mpsrch and BLAZE for the MasPar) and for new hardware (from Compugen and the NCBI), as well as IBM's offering called FLASH. Also note that BLAST, produced by NCBI, is a very common, rapid database homology search program. BLITZ (provided via an e-mail server: BLITZ@embl-heidelberg.de) is also very popular because it performs a full Smith-Waterman search of the database.

8. *GENEPRO* (DNA Gel Reader), (IBM PC, MS-DOS/PC-DOS, IBM PS/2, MS-DOS/PC-DOS), Riverside Scientific Enterprises, 18332 57th Ave., N.E., Seattle, WA 98155, USA.

9. *GENETYX, GENETYX-MAC* (nucleic acid and protein analysis), (Macintosh, Sun-3, Sun-4, Unix), Keiichi Niwa, Chief Manager, Biotechnology Division, Software Development Co. Ltd., 3-8-12 Shibuya, Shibuya-ku, Tokyo 150, Japan.

10. *GENEius* (sequence analysis software), (Apple II, GS/Pro-DOS, Macintosh), BioTechnology Software Specialities, P.O. Box 1521, Corvallis, OR 97339, USA.

11. *GeneWorks* (nucleic acid and protein analyses of all types), (Macintosh), Murray Summers, Director of Sales, IntelliGenetics Inc., 700 E. El Camino Real, Mountain View, CA 94040, USA.

12. *IG Suite* (nucleic acid and protein analyses of all types), (Sun-3 and Sun-4, Unix, Sun SparcStations, Unix, VAX, VMS), Murray Summers, Director of Sales, IntelliGenetics Inc., 700 E. El Camino Real, Mountain View, CA 94040, USA.

13. *LaserGene* and *DNASTAR* (nucleic acid and protein analyses of all types), (IBM PC, MS-DOS/PC-DOS, IBM PS/2, MS-DOS/PC-DOS, Macintosh), Jerry Miller, DNASTAR, Inc., 1228 S. Park St., Madison, WI 53715, USA.

14. *MALI and PRALI* (multiple alignment of proteins), (Sun, Unix, Vax), Martin Vingron, Biocomputing Programme, Postfach 10.2209, Meyerhofstrasse 1, Heidelberg D-6900, Germany.

15. *NBRF-PIR* (sequence analysis software), (VAX, VMS), Kathryn Sidman, Protein Identification Resource, National Biomedical Research Foundation, 3900 Reservoir Rd., N.W., Washington, DC 2007-2195, USA.

16. *PC/GENE* (nucleic acid and protein analyses of all types), (IBM PC, PS/2) IntelliGenetics, Inc., Murray Summers, Director of Sales, 700 E. El Camino Real, Mountain View, CA 94040, USA.

17. *PROSIS, PSQ* (protein sequence database retrieval system), (VAX, VMS), Kathryn Sidman, Protein Identification Resource, National Biomedical Research Foundation, 3900 Reservoir Rd., N.W., Washington, DC 20007-2195, USA.

18. *Prostruc* (secondary structure predictions), (computers running Fortran 77), Molecular Biology Computer Research Resource, Dana-Farber Cancer Institute, 44 Binney St., Boston, MA 02115, USA.

19. *Recombinant Toolkit 4.0* (nucleic acid and protein sequence analysis), (IBM PC), Dr. Stefan Unger, SOFTSHU, 2250 Webster, Palo Alto, CA 94301, USA.

20. *SEQAID II* (sensitive sequence comparison), (comparison of proteins for distance relationships), (VAX, SUN), R. Rechid, EMBL, Postfach 10.2209, Meyerhofstrasse 1, Heidelberg D-6900, Germany.

21. *DNA Strider for the Mac* (sequence analysis), (Mac), Dr. Christian Marck, Service de Biochimie et de Genetique Moleculaire, Bat. 142 Centre d'Etudes de Saclay, 91191 Gif-Sur-Yvette, Cedex France. Fax: (33 1) 69 08 47 12.

22. *Genetic Data Environment* (X-Windows), by Steve Smith and collaborators, smith@nucleus.harvard.edu.

23. *GCG*, produced by Genetics Computer Group, Inc., University Research Park, 575 Science Drive, Suite B, Madison, WI 53711, USA. Tel: (608) 231-5200, help@gcg.com (sequence analysis package).

For those readers interested in an example of software allowing users to perform a dissection, practice a surgical procedure, or help others understand anatomy, A.D.A.M. may be of interest. The fully interactive, multimedia medical computer software features on-screen dissection of the human anatomy. Some have said that this is the 21st century's answer to *Gray's Anatomy*. For further information, contact A.D.A.M. Software, 1899 Powers Ferry Road, Suite 460, Marietta, GA 30067, USA.

A Biologist's Guide to Internet Resources

A Biologist's Guide to Internet Resources, written by Una R. Smith (Yale University), is an invaluable guide to computers, databases, and biology. This guide is updated monthly. The most current version is available via Usenet, gopher, FTP, and e-mail. In Usenet, look in sci.bio or sci.answers. You may also FTP to rtfm.mit.edu. Give the username "anonymous" and your e-mail address as the password. Use the "cd" command to go to the directory pub/usenet/ news.answers/biology/ and use "get guide" to copy the file to your computer. You may also send e-mail to mail-server@rtfm.mit.edu with the text "send usenet/news.answers/biology/guide/*".

Una R. Smith's guide contains information on bibliographies, information archives, software, databases, and access tools.

Also included is information concerning on-line services, digests, and discussion groups, such as Artificial life digest, Biological Anthropology, Biological timing and circadian rhythms, Biologia y Evolucion, Biology information systems, Bulletin for bryologists, Cytometry discussion, Dendrome forest tree genome mapping digest, Dinosaurs and other archosaurs, Discover Insight Biosym Users' Group, Ecologia (in Spanish), Entomology discussion, Environmentalists digest, Fish and Wildlife Biology, Forestry discussion, Genstat statistics package discussion, GIS digest, GIS Users in the United Kingdom, Killifish, Cyprinodontidae, Neotropical birds discussion, Neural networks digest, Orchids, Plant Taxonomy, Primate discussion, Prion Research Digest,

Young Scientists' Network, Volcano list, Bean Bag: Leguminosae Research Newsletter, Botanical Electronic News (BEN), Environmental Resources

Information Network (ERIN) Newsletter, Climate/Ecosystem Dynamics (CED), and The Chlamydomonas Newsletter.

Genetics, Evolution, and Biology in Science-Fiction

"Science fiction, as a literature of change, of future lifeforms and alien life evolving beyond earth, is implicitly and deeply influenced by the theory of evolution."

James Gunn
The New Encyclopedia of Science-Fiction (1988)

Bishop, M., *No Enemy But Time* (Bantam, 1985).[20]

Engel, A., *Variant* (Donald Fine, 1988).[21]

Herbert, F., *The White Plague* (Berkley, 1983).[22]

Huxley, J., "The tissue culture king", *Time Probe: The Sciences in Science Fiction*, ed. A. Clarke (Dell, 1966).

Pickover, C., "Extraterrestrial messages in our genes", *Mazes for the Mind: Computers and the Unexpected* (St. Martin's Press, 1992).

Pickover, C., "Ghost children in our genes", *Mazes for the Mind: Computers and the Unexpected* (St. Martin's Press, 1992).

Harrison, H., *West of Eden* (Bantam, 1985).[23]

[20] A hominid survivor appears in modern Georgia.
[21] An excellent fictional account of gene manipulation with startling consequences for several young Russian boys.
[22] This book is about a man whose two daughters and wife are blown up by an IRA bomb. The man develops a selective virus which is carried by males and kills females, and sends it to Ireland.
[23] *Saurus sapiens* dinosaurs evolve along with *Homo sapiens* and live together.

Herbert, F., *Eyes of Heisenberg* (Berkley, 1986).[24]

Wells, H., "The man of the year million" (1893).[25]

Hamiltion, E., *The Best of Edmond Hamilton* (Ballantine, 1977).[26]

Stableford, B., *The Gates of Eden* (Daw Books, 1983).[27]

Stapledon, O., *Last and First Men* (Tarcher, 1930).[28]

Watson, I., *The Martian Inca* (Ace, 1977).[29]

Silverberg, R., *Son of Man* (Ballantine, 1971).[30]

Sheffield, C., *Sight of Proteus* (Ace, 1978).[31]

Sterling, B., *Schismatrx* (Ultramarine, 1985).[32]

Galouye, D., *Project Barrier* (Gollanczc, 1968).[33]

Brin, D., *Startide Rising* (Bantam, 1983).[34]

Lem, S., *The Invincible* (Ace, 1964).[35]

Williamson, J., *Dragons Island* (Tower, 1951).[36]

[24]Genetic engineering of immortal men.
[25]The evolution of humans toward human tadpoles with "a dangling degraded pendant to their minds".
[26]See "The man who evolved" and "Devolution", where all terrestrial life is an example of degeneration from superior, shape-shifting jelly-beings.
[27]An alien species has separate genetic codes for different body forms. Describes a metamorphic multispecies which can absorb and express human DNA.
[28]Two billion years of human future evolution. Includes bioengineering and evolutionary cul-de-sacs.
[29]Martian aggregating slime-molds rewire human brains.
[30]Genetic engineering.
[31]Genetic engineering.
[32]Genetic engineering.
[33]Higher mammals, such as dogs and bears, are allowed to evolve language, technology, and social structures.
[34]Genetic engineering of dolphins.
[35]A collective cloud of small cybernetic crystals evolves and destroys organic life.
[36]Controlled mutation and the creation of "non-men".

Bell, E., "Seeds of life", *Amazing Stories Quarterly* (Dunellen, 1931) **4**(4), 434, 505, 520 (illus.).[37]

Sturgeon, T., "Microcosmic God", *Without Sorcery* (Prime Press, 1941).[38]

Blish, J., *The Seedling Stars* (Gnome Press, 1957).[39]

Benford, G., *In the Ocean of Night* (Bantam, 1977).[40]

Bear, G., *Blood Music* (Ace, 1985).[41]

Crichton, M., *Jurassic Park* (Ballentine, 1990).[42]

> *"What is written in science fiction today may in the future be written into our genes."*

James Gunn
The New Encyclopedia of Science-Fiction (1988)

Acknowledgment

Several participants of the bioforum newsgroup on the usenet computer network provided useful information for this preface and I thank them for the advice and comments.

---

[37]Evolution.
[38]Making life in the laboratory.
[39]The alteration of human beings for survival in alien environments.
[40]Biology's evolution toward mechanical, computerized forms.
[41]Computer circuits based on organic molecules. Miniature intelligence.
[42]Describes the cloning of dinosaurs from fossil DNA.

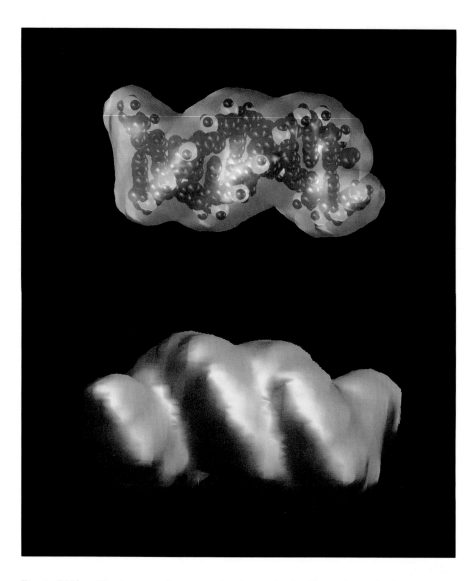

Fig. 1. DNA, with *shape-envelope* computed to emphasize the major and minor grooves. Rather than looking at all the atoms in a biomolecule, it is sometimes easier to obtain a visual feel for the shape by plotting a 3-D isosurface using an $1/r_i^4$ dependence, where r is the distance of each atom to a grid element. For details see: Pickover, C., *Computers and the Imagination* (St. Martin's Press, 1992). (Chapter on "virtual voltage sculpture".)

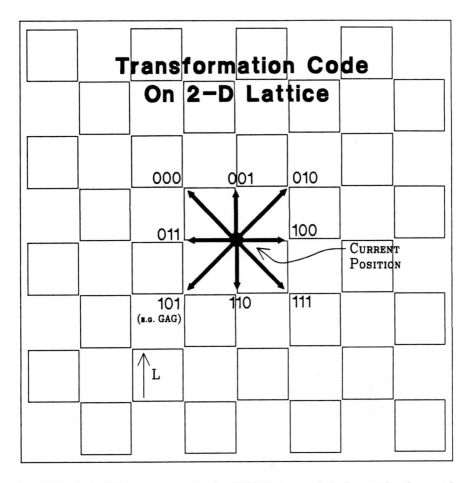

Fig. 2. Visualizing DNA sequences using the "DNA Vectorgram". In the past decade, several researchers have converted the sequence of DNA bases to characteristic patterns traced out on a lattice in order to visually locate biologically relevant features. In the vectorgram transformation shown here, DNA is represented as a string of 0's and 1's, and overlapping triplets of digits cause the graphics "pen" to travel to nine different positions on a lattice. Using this method, various structural themes and repeats are easy to locate. (G=1, C=1, A=0, T=0.) For details see: Pickover, C., *Computers, Pattern, Chaos, and Beauty* (St. Martin's Press, 1991). Also: Pickover, C., "DNA vectorgrams: Representation of cancer gene sequences as movements along a 2-D cellular lattice", *IBM Journal of Research Development* **31**, 111–119 (1987).

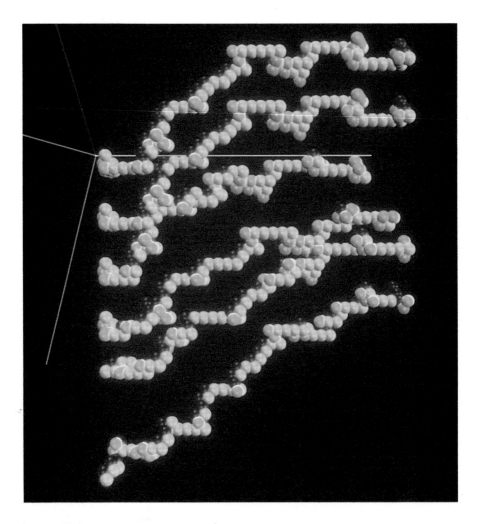

Fig. 3. DNA and amino acid tetragrams. Both DNA and amino acid sequences can be represented by inspecting the sequence, one base or residue at a time, and assigning a tetrahedral direction of movement corresponding to the base. For amino acids, the four directions correspond to four categories of amino acids: Polar, nonpolar, positively charged, and negatively charged. Shown here is an evolutionary sequence of amino acid tetragrams for a cytochrome C sequence from a human (top), spider monkey, penguin, rattlesnake, honeybee, and spinach (bottom). As expected, more closely related organisms give similar tetragrams. For details, see: Pickover, C., "DNA and protein tetragrams: Biological sequences as tetrahedral movements", *Journal of Molecular Graphics* **10**(1), 2–6 (1992).

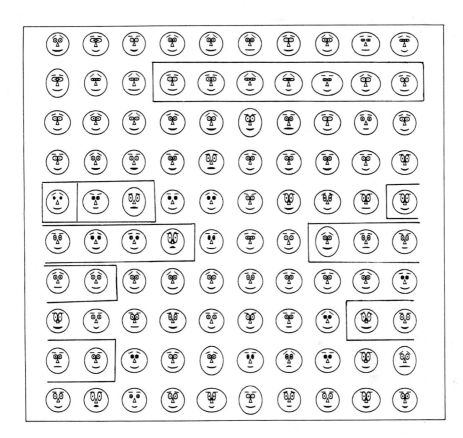

Fig. 4. Computer-drawn faces used as a multidimensional representation to represent statistical properties of the sequence of bases in the DNA of a human bladder cancer gene. The deviation of ten statistical properties of the DNA are mapped to deviations in facial coordinates from their middle settings. Facial symbols are more reliable and memorable than other icons and allow the human analyst to visually grasp many of the essential regularities and irregularities in the data. For more details, see: Pickover, C., "Computer-drawn faces characterizing nucleic acid sequences", *Journal of Molecular Graphics* **2**, 107–110 (1984).

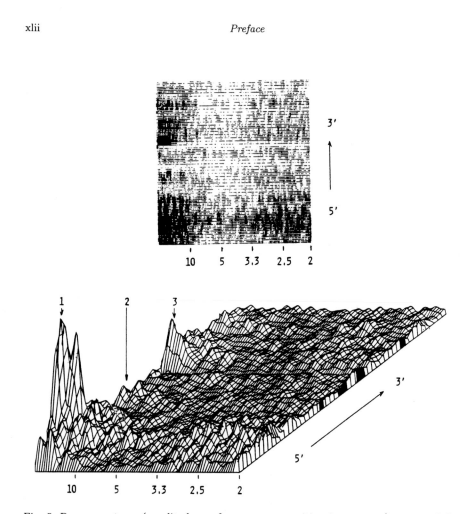

Fig. 5. Power spectrum (amplitude vs. frequency vs. position in sequence) computed for the DNA sequence of a human bladder cancer gene. The DNA sequence may be thought of as running from front to back (5' to 3' end). Four exons (1670-1779), (2047-2226), (2381-2540), (3238-3354) are indicated by shaded side regions. Frequency of G-C to A-T changes are represented on the x axis (low frequency on the left), with period values actually labeled to facilitate interpretation (see text). The three peaks (1, 2, 3) indicated by arrows are discussed in: Pickover, C., "Frequency representations of DNA sequences: Application to a bladder cancer gene", *Journal of Molecular Graphics* **2**, 50 (1984). Also: Pickover, C., *Computers, Pattern, Chaos, and Beauty* (St. Martin's Press, 1991). The spectrogram (inset), an alternate representation of the same data, portrays position in sequence (ordinate) vs. frequency (abscissa). Amplitude is indicated by darkness on the plot.

List of Contributors

Rosmarie Swanson and Stanley M. Swanson
Department of Biochemistry and Biophysics
Texas A&M University
College Station, TX 77843-2128
USA

Ann Williams, Kelly D. Chenault and Ulrich Melcher
Department of Biochemistry
Oklahoma State University
Stillwater, OK 74078-0454
USA

Peter K. Rogan
Department of Paediatrics
Milton S. Hershey Medical Center
Pennsylvania State University
Hershey, PA 17033
USA
E-mail: rogan@ncifcrf.gov

Joseph J. Salvo
Division of Biological Sciences
GE-Corporate Research and Development
Schenectady, NY 1230-0008
USA
E-mail: salvo@crd.ge.com

R. Michael Stephens
National Cancer Institute
Frederick Cancer Research and Development Center
Laboratory of Mathematical Biology
Ijamsville, MD 21754
USA

Thomas D. Schneider
National Cancer Institute
Frederick Cancer Research and Development Center
Laboratory of Mathematical Biology
P O Box B
Frederick, MD 21702-1201
USA
E-mail: toms@ncifcrf.gov

Jacques Ninio
Ecole Normale Superieure
Laboratoire de Physique Statistique
24 rue Lhomond
75231 Paris cedex 05
France
E-mail: ninio@physique.ens.fr

Eduardo Mizraji
Facultad de Ciencias
Universidad de la Republica
Tristan Narvaja 1674
11200 Monterideo
Uruguay
E-mail: mizraj@fcien.edu uy

Peter F. Lemkin
Image Processing Section
Laboratory of Mathematical Biology
Division of Cancer Biology, Diagnosis and Centers
National Cancer Institute
Frederick Cancer Research and Development Center
P O Box B
Buliding 469, Room 150B
Frederick, MD 21702-1201
USA
E-mail: lemkin@ncifcrf.gov

D. A. Kuznetsov
Engelhardt Institute of Molecular Biology
Rsuuian Academy of Sciences
Vavilov Street 32
117 984 Moscow B-334
Russia

H. A. Lim
Supercomputer Computations Research Institute
400 Science Center Library
Florida State University, B-186
Tallahassee, FL 32306-4052
USA
E-mail: hlim@scri.fsu.edu

Nobuo Munakata
Radiobiology Division
National Cancer Center Research Institute
Tsukiji 5-1-1, Chuo-ku
Tokyo 104
Japan
E-mail: nmunakat@ncc.go.jp

Kenshi Hayashi
Institute of Experimental Gene Informatics
Kyushu University
3-1-1, Maidashi
Fukuoka 812
Japan
E-mail: khayashi@gen.kyushu-u.ac.jp

Chun-Ting Zhang
Department of Physics
Tianjin University
Tianjin 300072
China

Jose Campione-Piccardo
National Laboratory for Viral Oncology
Laboratory and Center for Disease Control
Tunney's Pasture, Virus Building
Ottawa, Ontario
Canada K1A 0L2

Perry B. Hackett
Department of Genetics and Cell Biology
250 Biological Sciences Center
University of Minnesota
1445 Gortner Avenue
St. Paul, MN 55108-1095
USA
and
Institute of Human Genetics
University of Minnesota
Minneapolis, MN 55455
USA
E-mail: perry@molbio.cbs.umn.edu

Mark W. Dalton
Cray Research, Inc.
CIC-7/CON, MS B294
Los Alomos National laboratory
Los Alamos, NM 87545
USA
E-mail: mark@alamos.cray.com

Darrin P Johnson
Cray Research,Inc.
Cray Research Park
655F Lone Oak Drice
Eagan, MN 55121-1560
USA
E-mail: darrin@cray.com

Melvin R. Duvall
Department of Genetics and Cell Biology
University of Minnesota
St. Paul, MN 55108-1095
USA
and
Department of Botany and Plant Sciences
University of California
Riverside, CA 92521
USA

Michel T Semertzidis, Eitienne Thoreau, Anne Tasso, and Jean Paul Mornon
Systemes Moleculaires et Biologie Structurale
Universites Pierre et Marie Curie (Paris VI) et Denis Diderot (Paris VII) P7
Laboratoire de Mineralogie-Cristallographie
CNRS URA 09
Tour 16 - Case 115
4 Place Jessieu
F-75252 Paris cedex 05
France

Bernard Henrissat
Centre de Recherches surles Macromolecules Vegetales
CNRS
Universite Joseph Fourier
B P 53X
F-38041 Grenoble
France

Isabelle Callebaut
Unite de Biologie Moleculaire et Physiologie Animale
Faculte de Sciences Agronomiques
Passage des Deportes
2, B-5030 Gembloux
Belgique

T. K. Attwood
Department of Biochemistry and Molecular Biology
University College London
Gower Street
London WC1E 6BT
UK
E-mail: attwood@bsm.biochemistry.ucl.ac.uk

David Parry-Smith
Department of Molecular Sciences
Pfizer Central Research
Sandwich
Kent CT13 9NJ
UK

Kenji Yamamoto
Laboratory of Clinical microbiology and Immunology
Bun'inn Hospital
University of Tokyo
3-38-6, Mejirodai
Bunkyo-ku
Tokyo 112
Japan
E-mail: backen@camille.is.s.u-tokyo.ac.jp

Hiroshi Yoshikura
Department of Bacteriology
Faculty of Medicine
University of Tokyo
7-3-1 Hongo
Bunkyo-ku
Tokyo 113
Japan

Y. K. Huen
CAH Research Centre
P O Box 1003
Kent Ridge Post Office
Singapore 0511
Singapore
E-mail: chehyk@nusvm.nus.sg

Contents

Preface   vii

List of Contributors   xliii

A Picture of the Genetic Code   1
  R. Swanson and S.M. Swanson

Graphic Representations of Amino Acid Sequences   6
  A. Williams, K.D. Chenault and U. Melcher

Representing Protein Sequence and Three-Dimensional Structure
in Two Dimensions   15
  R. Swanson

Visual Display of Sequence Conservation as an Aid to Taxonomic
Classification Using PCR Amplification   21
  P.K. Rogan, J.J. Salvo, R.M. Stephens and T.D. Schneider

Perceptible Features in Graphical Representations of Nucleic
Acid Sequences   33
  J. Ninio and E. Mizraji

Representations of Protein Patterns from Two-Dimensional
Gel Electrophoresis Databases   43
    P.F. Lemkin

A Protein Visualization Program   61
    D.A. Kuznetsov and H.A. Lim

Gene Music: Tonal Assignments of Bases and Amino Acids   72
    N. Munakata and K. Hayashi

Diagrammatic Representation of Base Composition in DNA
Sequences   84
    C.-T. Zhang

A Transforming Function for the Generation of Fractal Functions
from Nucleotide Sequences   96
    J. Campione-Piccardo

Visualization of Open Reading Frames in mRNA Sequences   119
    P.B. Hackett, M.W. Dalton, D.P. Johnson and M.R. Duvall

Visualization of Protein Sequences Using the Two-Dimensional
Hydrophobic Cluster Analysis Method   129
    M.T. Semertzidis, E. Thoreau, A. Tasso, B. Henrissat,
    I. Callebaut and J.P. Mornon

Diagnosis of Complex Patterns in Protein Sequences   145
    T.K. Attwood and D.J. Parry-Smith

RNA Folding and Evolution   158
    K. Yamamoto and H. Yoshikura

Representation of Biological Sequences Using Point Geometry
Analysis   165
    Y.K. Huen

Index   183

A Picture of the Genetic Code

Rosemarie Swanson
and
Stanley M. Swanson

A cell's working knowledge of the genetic code is embodied in its tRNA syn-
thases — the enzymes that charge the codon-recognizing tRNA molecules with
their appropriate amino acids. However, the cell's use of the tRNA molecule
as a physical tie between a codon and an amino acid could in principle have
freed the genetic code from any restrictions on which amino acid goes with
which codon. The relation between the nucleotide "spelling" of a codon and
the properties of the amino acid it specifies could be random. Such is **not**
the case, however. Codons are "spelled" by nucleotides in a way that reduces
the impact of reading errors. A simple relation exists between the spelling of
codons, the physical properties of the amino acids, and the interchangeabil-
ity of amino acids in proteins. The relation is summarized graphically, with
the genetic code arranged as a Gray code and the amino acid characteristics
correlated with the Gray code structure.

Error control is near the heart of good technological code design, and there
is every reason to assume that error effects were also of utmost significance
in the evolution of the genetic system. The genetic code is known to have
a structure that reduces the impact of errors in converting RNA code into
proteins.[1] Another example of such an error-reducing structure also exists in

ideal form in a simple mathematical code — a Gray code[2] (see Glossary for an example). The error reduction in a Gray code is a consequence of its minimum change design, a design in which look-alike "words" are assigned similar "meanings" so that the impact of the most likely kind of reading error is minimized. Code words which are less likely to be confused with each other by misreading are assigned to meanings that are more important to distinguish.

Is there a simple way to derive and express the "meaning" relationships among the amino acids? The "meaning" of an amino acid is its role in determining the three-dimensional structure of proteins. The best measure of similarity of meaning is the ability of one amino acid to substitute for another in a similar protein structure. Data[3] from 1572 substitutions in 71 groups of closely related sequences were reduced[1] to an ordering of the amino acids around a circle, in which neighboring amino acids have high probabilities of replacing each other, and more distant amino acids are less likely replacements. The ordering was objective, based solely on replacement probabilities, with no consideration of other characteristics. However, the resulting ordering concentrated more hydrophilic (likely to be on the surface of the protein in contact with water) amino acids in one half of the circle and more hydrophobic (likely to be on the inside of a protein, sequestered from water) amino acids in the other half. If these halves are chosen as top and bottom, then the *right* and *left* halves of the circle show a clean division into larger and smaller amino acids, respectively.

The ordering of the amino acids according to their protein substitution similarity is shown around the innermost circle in Fig. 1. (Both compact [one-letter] and mnemonic [three-letter] abbreviations for the amino acids are used. The figure caption lists their full names). Just inside the innermost circle, the groupings that contain the largest, the smallest, the most hydrophilic, and the most hydrophobic amino acids are indicated by labeled arcs. For comparison, in the central area of the figure is a scatter plot[1] of the amino acid physical properties of side chain volume (abscissa) and hydrophobic free energy (ordinate), from which can be seen a good correlation between the substitution similarity order and the physical properties. Thus, the amino acids show a clear and simple organization with respect to substitution similarity.

Turning to the genetic code again, there is an equally simple arrangement of codons according to their nucleotide spelling that translates to an amino acid ordering with striking similarities to the natural substitution ordering. This codon arrangement is shown in the outer three rings of Fig. 1. The

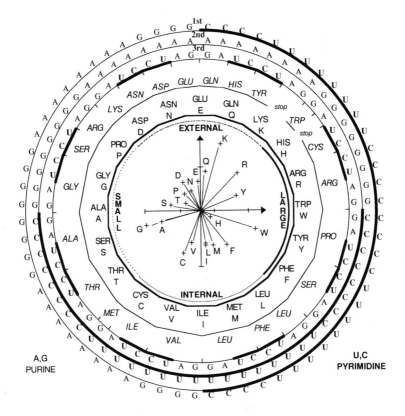

Fig. 1. The genetic code is arranged in Gray code format according to reference 1 and compared with a substitution-based similarity ring for amino acids and a graph of their measured physical properties. Volume, horizontal axis; hydrophobic free energy, vertical axis. The axes are not labeled, but the center of the plot is at the median values of the two properties, 43 Å^3 and 0.1 kcal/mole. The tick marks on the volume axis are at 0, 25, 50, and 75 Å^3, and at -1.0 and 1.0 kcal/mole on the energy axis. U – uracil, C – cytosine, G – guanine, A – adenine; ala – alanine, thr – threonine, gly – glycine, pro – proline, ser – serine (small, common amino acids — comprise 35% of the population, i.e., 35% of the total number of amino acids in 314 representative sequences for 314 families of proteins[3]); asp – aspartic acid, asn – asparagine, glu – glutamic acid, gln – glutamine, lys –lysine (hydrophilic amino acids, on average about four times as likely to be on the surface of a protein as the small residues — compose about 25% of the population); his – histidine, arg – arginine, trp – tryptophan, tyr – tyrosine, phe – phenylalanine (large, less common, π-electron containing amino acids — compose about 15% of the population); leu – leucine, met – methionine, ile – isoleucine, val – valine, cys – cysteine (hydrophobic, about four times as likely to be on the inside of proteins as the small amino acids — compose about 25% of the population). Population percentages derived from reference 3, Suppl. 3.

second nucleotide "letter" of each codon is a purine (A or G) in the top half of the diagram and is a pyrimidine (U or C) in the bottom half, corresponding to the hydrophilic-hydrophobic division in the amino acid substitution circle. Likewise, the first nucleotide "letter" of each codon is a purine in the left half of the diagram and is a pyrimidine in the right half, corresponding to the small-large division in the substitution circle. This pattern corresponds to the simplest possible Gray code. The translations of the codons to amino acids are shown in the polygon just inside the three outer codon rings. The ordering of the amino acids according to the translation clearly shows the general trends of hydrophilic-hydrophobic top-bottom division and large-small right-left division, even though there are some anomalous translations (pro and cys most notably) and some "misplaced, extra" codons (ser and arg).

The major divisions, top-bottom, right-left, of the genetic code diagram correspond to both the most important distinctions between amino acids and the most accurately readable "letters" of the codons. The translation machinery for the genetic code is very accurate at this point in evolution, superseding these very simple error reducing properties of the genetic code, but their interesting imprint remains, both in codon assignments and in the nucleotide sequences that code for proteins.

References

1. Swanson, R., "A unifying concept for the amino acid code", *Bull. Math. Biol.* **46**, 187–203 (1984) (and references therein).
2. Flores, I., *Encyclopedia of Computer Science*, eds. A. Ralston and C.L. Meek (Petrocelli/Charter, 1976) pp. 210–215.
3. Dayhoff, M.O., *Atlas of Protein Sequence and Structure* (National Biomedical Research Foundation, 1972–1978) Volume 5 and supplements 1–3.

Glossary

Gray code: Gray code is a code designed to be machine-readable with reduced error levels. It is a type of binary code in which similar bit patterns are used to designate similar values. "Similar" bit patterns are ones that differ in few positions and are most likely to be mistaken for each other by the kind of machine that reads the patterns. Thus, even if a bit pattern is misread, the resulting error is most likely to be small. An example of a

Gray code using three binary digits to encode an angular position is

Gray code	111,	110,	100,	101,	001,	000,	010,	011,	111...
representing	0,	$\pi/4$,	$2\pi/4$,	$3\pi/4$,	π,	$5\pi/4$,	$6\pi/4$,	$7\pi/4$,	2π...

Notice that adjacent triples differ at only one position, and differ most frequently in the last position. For further information, see Gardner, M., "Mathematical games", *Scientific American* **227**(2), 106–109 (1972) and Heath, F.G., "Origins of the binary code", *ibid.* 76–83.

Graphic Representations of Amino Acid Sequences

Ann Williams, Kelly D. Chenault
and
Ulrich Melcher

Proteins and nucleic acids are linear biological polymers of 20 and 4 building blocks respectively. The sequence of building blocks along the polymer determines the properties of the protein or nucleic acid. By convention, letters of the Roman alphabet represent such sequences. This representation consumes more space than necessary and hampers the visual recognition of interesting and important features of the sequences. The Puppy representation of sequences of nucleotides overcomes these problems for nucleic acids. We designed two representations for the sequences of amino acid residues in proteins. In a minimal representation, each residue type is depicted by a different series of vertically aligned marks. In the second representation, "Kitty", the characters reflect chemical structures of the residues. Comparison of the two representations for the amino acid sequences of three proteins suggested that the "Kitty" method may be more useful in displays of sequences.

Proteins are linear biological polymers constructed from 20 types of amino acids. The sequence of amino acid residues is usually represented by an N-terminal to C-terminal string of abbreviations. These are either three-letter or, more commonly, one-letter[1] abbreviations. Because the names of several residues begin with the same letter, some one-letter codes bear no obvious relation to the name of the amino acid they represent (e.g., Q for glutamine, W for tryptophan, and K for lysine). The visual appearance of the characters of the Roman alphabet that is used for these codes bears no relation to the structures or chemical properties of the residues they represent and can, in some fonts, be confused for other characters (e.g., G for C, V for Y, and uppercase I for lower case l). Yet, the one-letter code is more space-efficient than the three-letter abbreviations.

Only four residue types, also represented by single letters of the Roman alphabet, make up the building blocks of each of the nucleic acids. Several methods, alternatives to use of Roman characters, have been proposed to represent the sequence of nucleotides in these linear biological polymers.[2,3] In the Puppy representation,[3] named for purines and pyrimidines (the two types of chemical bases represented), nucleotides are represented by three vertically aligned spaces. An occupied, lowest space denotes a pyrimidine, an occupied uppermost space a purine; occupation of the middle position indicates a guanine or cytosine base. The representation is efficient in its use of space and allows visual recognition of many patterns important to the biological functions of the nucleic acid.

Alternate ways of displaying amino acid sequences have received considerably less attention. Values for various chemical properties of regions within a polypeptide can be plotted as a function of position along the chain.[4] Typically, the graphs display hydrophobic or hydrophilic tendencies and probabilities of forming various types of secondary structure. The graphs display the average character of groups of contiguous residues. Reducing the size of the groups to one residue results in a seemingly erratic plot that obscures trends in the chemical properties. Two groups[5,6] proposed a two-dimensional vectorial display of amino acid residue sequences. Residue-specific vectors were derived from the likelihood of substitution of one residue for another.[7] This representation proved useful in identifying regions of sequence similarity in different proteins.[8,9] Yet, it does not allow ready identification of individual amino acid residues. As an aid to the alignment of the sequences of multiple related proteins, Attwood *et al.*[10] used color to aid in the recognition of residues and residue types. Yet, the use of color is a luxury not affordable by authors of many journal presentations and is not available on all computer terminals. Further, the human memory for colors is worse than it is for patterns.

Representations of amino acid sequences that overcome the deficiencies of current methods should satisfy several criteria: They should be efficient in their use of space; they should allow the facile and unambiguous recognition of individual residue types; and at the same time, they should convey a sense of the properties of the polypeptide chain in which they are embedded. We designed several new representations, of which we present the two most promising here. The first is a minimal representation and the second is based on the chemical structures of the amino acids. To test and illustrate the methods, we chose proteins that represent examples of the three major types of protein secondary structure: β structure, α helical rods, and the triple helix. The sequences chosen were obtained from a databank of protein sequences (Swiss-Prot) and had accession numbers P01859 (human immunoglobulin γ-2 chain constant region), P02564 (rat cardiac muscle β isoform of myosin heavy chain), and P20849 (human collagen a1(IX) chain precursor). The representations were created with an Apple IIe computer using the font editor of the Multiscribe word processing program (version 2.00, Styleware, Inc.). The fonts are available on request from the corresponding author.[*] A Macintosh truetype and postscript versions of the "Kitty" font, combined with the "Puppy" font,[3] are available as pupkit.hqx by anonymous ftp from the EMBL software server (ftp.embl-heidelberg.de) in the directory: /pub/software/Mac.

Representation of the 20 different residue types commonly encoded by nucleic acid templates requires a minimum of five binary bits. Five positions

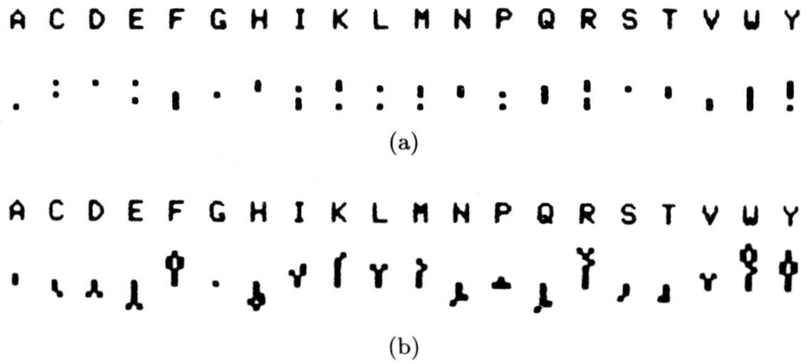

(a)

(b)

Fig. 1. Symbols used in amino acid sequence representations. Amino acid designations are given in the one-letter code: A — alanine; C — cysteine; D — aspartic acid; E — glutamic acid; F — phenylalanine; G — glycine; H — histidine; I — isoleucine; K — lysine; L — leucine; M — methionine; N — asparagine; P — proline; Q — glutamine; R — arginine; S — serine; T — threonine; V — valine; W — tryptophan and Y — tyrosine. (a) Minimal representation and (b) structural "Kitty" representation.

[*]Ulrich Melcher

provide 32 possible symbol types. In the minimal representation, we arranged five spaces vertically and filled in the spaces according to a scheme that attempts to take account of the chemical properties of the residues (Fig. 1(a)). The top space is occupied if the residue is potentially charged (cys, asp, glu, his, lys, arg, tyr). The next lower space is occupied if the residue can form a hydrogen bond or if it is a long aliphatic amino acid (his, ile, lys, leu, met, asn, gln, arg, ser, thr, tyr). The acidic residues and cys are excluded from this group. The middle space is occupied for gly, pro, thr, and residues containing sulfur atoms, aromatic residues or amides (cys, phe, met, asn, gln, trp, tyr). The next to lowest space is filled in for phe, ile and trp and the larger of four pairs of similar residues (glu, gln, arg, val). The bottom space is occupied for residues that are hydrophobic, including the basic residues (ala, phe, ile, lys, leu, met, pro, arg, val, trp, tyr). Hereinafter, horizontally contiguous spaces are referred to as rows and the collection of five rows that represent the sequence, a sequence line.

Figure 2 shows the amino acid sequences of three proteins in the minimal representation. The constant region of Ig γ heavy chains (Fig. 2(a)) consists of three domains of similar structure and size. A proline-rich hinge region separates the first domain from the second. The hinge region appears in the minimal representation as interrupted lines in the middle and bottom rows (indicated by "h" in Fig. 2(a)). A similar, but shorter, sequence separates the last two domains. Immunoglobulin domains consist of β sheets, many of which have a hydrophobic and a hydrophilic face. The resulting alternation of hydrophobic and hydrophilic residues gives a dotted appearance to some sections of the lowest row of the sequence lines (for example, at "b" in Fig. 2(a)). The collagen chain (Fig. 2(b)) has an N-terminal globular domain and three triple-helical domains. The triple-helical domains, consisting of pro.gly.X repeats, where X represents any amino acid (but is often pro), show up clearly as solid center rows and a dashed lower row (indicated by "p" in Fig. 2(b)). As a result, the N-terminal domain and the two smaller regions that separate triple-helical domains were easily spotted. Myosin (Fig. 2(c)) also consists of a globular N-terminal domain and a rod-shaped C-terminal domain. In the minimal representation, the transition between globular and rod domains appeared as a shift (indicated by "r" in Fig. 2(c)) from a diversity of residues to a deficiency of residues that call for a mark in the middle row. The rod-shaped domain forms an α helical coiled coil and has repeating motifs characterized by aliphatic residues (principally leu) at positions n, $n + 3$, and $n + 7$. These

(a)

(b)

(c)

Fig. 2. Amino acid sequences of selected proteins in the minimal representation (Fig. 1(a)): (a) Immunoglobulin γ chain constant region; (b) collagen and (c) myosin heavy chain. Letters indicate features mentioned in text: β structure example — b; α structure example — a; hinge-like region — h; beginning of triple-helical region — p and beginning of rod-shaped region — r.

repeating motifs (for example, at "a" in Fig. 2(c)) are not readily apparent in the minimal representation.

The structural representation is based directly on structures of the amino acid residues (Fig. 1(b)). The arrangement of pixels for each residue type closely approximated the number and connectivity of carbon, oxygen, nitrogen and sulfur atoms in the residues. The α carbon was included to allow marking of glycine. The "hydrophobic" amino acids were depicted as extending upward from the α carbon row, while the "hydrophilic" residues extended downward. Wherever possible the heteroatoms were placed either to the left or right of center. To distinguish serine from cysteine, placement of the mark for the heteroatom was to the left for the former and to the right for the latter. To distinguish acids from amides, the two oxygen atoms of acids were placed at the same horizontal level, but the nitrogen of amides was placed one position closer to the α carbon row. Proline was arbitrarily represented as three consecutive pixels in the α carbon row with one pixel in the row above. For simplicity, a bond closing the five-membered ring in tryptophan was omitted.

Figure 3 shows the three polypeptide sequences in the structural representation. Regions poor in bulky residues, or conversely, rich in bulky residues were easily recognized. As a result, the hinge region of the immunoglobulin γ heavy chain (indicated by "h" in Fig. 3(a)) was readily apparent. Alternating hydrophilic-hydrophobic sequences, frequently found in immunoglobulin β structure, had the appearance of characters separated by punctuation marks (for example, at "b" in Fig. 3(a)). The triple-helical domains of the collagen chain (beginnings indicated by "p" in Fig. 3(b)) also stood out due to their dearth of bulky residues. The presence of pro.gly.X repeats aided this recognition. As a result, the N-terminal globular domain and the regions separating triple-helical domains were also apparent. A visual scan of the myosin representation (Fig. 3(c)) did identify numerous regions in the C-terminal half where the branched chain aliphatic residues (val, leu, ile) occurred with spacings of alternately three and four residues (for example, at "a"). Such is the spacing expected of hydrophobic residues in molecules that form α helical coiled coils. In addition, many clusters of four to five predominantly basic residues and of two to three acidic or amide residues were also noted.

Both representations allow the identification of individual amino acid residues in the polypeptide chain, while conveying a sense of the overall sequence of the chain. Domains of different sequences can be recognized using either representation, as can smaller regions of specific sequence motifs. The

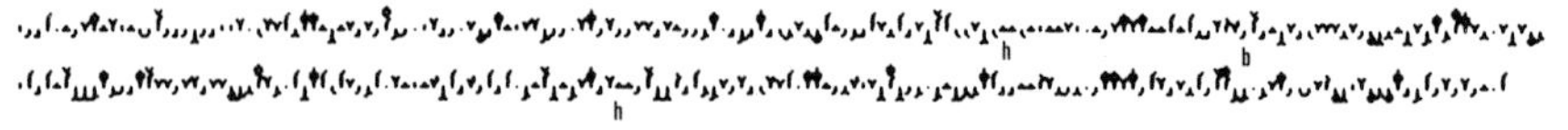

(a)

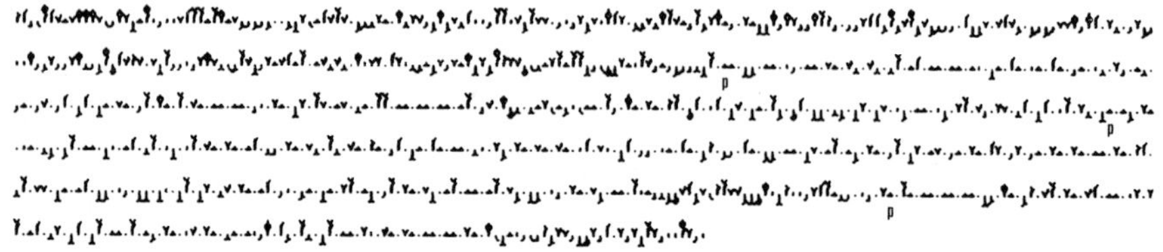

(b)

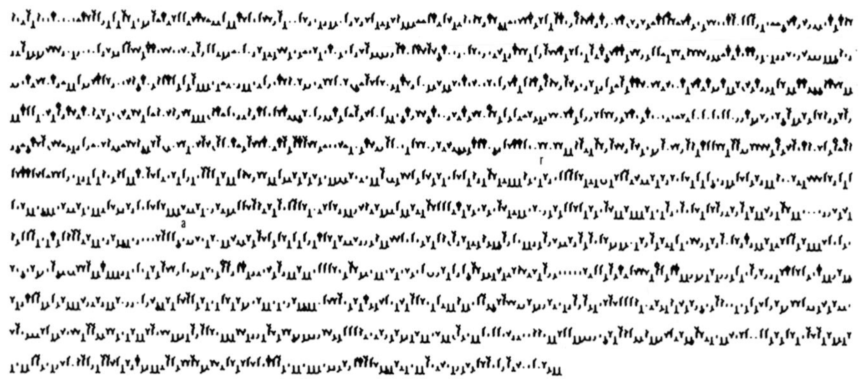

(c)

Fig. 3. Amino acid sequences of selected proteins in the "Kitty" representation (Fig.1(b)): (a) Immunoglobulin γ chain constant region; (b) collagen and (c) myosin heavy chain. Letters indicate features mentioned in text: β structure example — b; α structure example — a; hinge-like region — h; beginning of triple-helical region — p and beginning of rod-shaped region — r.

structural representation, measuring 12 by 3 pixels, is not as space-efficient as the minimal representation (5 by 1 pixel). Yet, the significance of the symbols in the structural representation is easily learned by individuals with knowledge of amino acid structures. Considerably greater effort is required to master the significance of the minimal pattern bars. We designate the structural representation of protein sequences as the "Kitty" representation since it is comparable in style and ease of residue recognition to the "Puppy" representation of nucleic acid sequences.

"Kitty" representations can be used in several ways. They could be used as a visual aid to the identification of the locations of sequence domains within a polypeptide sequence. Comparisons of "Kitty" representations of two or more proteins could serve as a useful adjunct to computer-assisted searches for amino acid sequence similarity. The representation may lead to the identification of hitherto unrecognized patterns in the distribution of residues and the elucidation of the biochemical significance of those patterns. Finally, if used in presentations of research results, the "Kitty" representation could provide a refreshing and useful way to convey information about an amino acid sequence.

Acknowledgements

Financial support from the Robert Glenn Rapp Foundation (to KDH), the Oklahoma Health Research Program and the Oklahoma Agricultural Experiment Station is gratefully acknowledged.

References

1. Dayhoff, M.O., *Atlas of Protein Sequence and Structure*, Vol. 5 (National Biomedical Research Foundation, 1972).
2. Hamori, E., "Graphic representation of long DNA-sequences by the method of H-curves: Current results and future aspects", *Biotechniques* **7**(7), 710–720 (1989).
3. Melcher, U., "A readable and space-efficient DNA sequence representation: Application to caulimoviral DNAs", *Comput. Appl. Biosci.* **4**(1), 93–96 (1988).
4. Doolittle, R.F., *Of URFs and ORFs* (University Science Books, 1986).
5. Poch, O., de Marcillac, G.D., Exinge, F., Roy, A. and Losson, R., "Functional domains of the regulatory protein PPR1: Use of the V. R. P. computer program", *Yeast* **4**(S), 416 (1988).
6. Swanson, R., "A vector representation for amino acid sequences", *Bull. Math. Biol.* **46**(4), 623–639 (1984).
7. French, S. and Robson, B., "What is a conservative substitution?", *J. Mol. Evol.* **19**(1), 171–175 (1983).
8. Fuchs, M., Pinck, M., Serghini, M.A., Ravelonandro, M., Walter, B. and Pinck, L., "The nucleotide sequence of satellite RNA in grapevine fanleaf virus, strain F13", *J. Gen. Virol.* **70**(4), 955–962 (1989).
9. Serghini, M.A., Fuchs, M., Pinck, M., Reinbolt, J., Walter, B. and Pinck, L., "RNA2 of grapevine fanleaf virus: Sequence analysis and coat protein cistron location", *J. Gen. Virol.* **71**(7), 1433–1441 (1989).

10. Attwood, T.K., Eliopoulos, E.E. and Findlay, J., "Multiple sequence alignment of protein families showing low sequence homology: A methodological approach using database pattern-matching discriminators for G-protein-linked receptors", *Gene* **98**(2), 153–159 (1991).

Glossary

pyrimidine: (abbreviation: Py) A class of nitrogenous heterocyclic bases, of which cytosine, thymine, and uracil are important constituents of nucleic acids (RNA and DNA).

purine: (abbreviation Pu) A class of nitrogenous heterocyclic bases, of which guanine and adenine are important constituents of nucleic acids (RNA and DNA).

hydrophobic: The property of certain amino acid residues of proteins that causes an increase in free energy upon their transfer from a non-polar solvent to water.

hydrophilic: The property of certain amino acid residues of proteins that causes a decrease in free energy upon their transfer from a non-polar solvent to water.

β structure: An extended conformation of a polypeptide backbone such as that found in a β sheet.

β sheet: An element of the three-dimensional structure of proteins that consists of two or more stretches of polypeptide backbone in an extended conformation and hydrogen bonded to one another.

α helical rod: An element of the three-dimensional structure of proteins that consists of a helical arrangement of one stretch of a polypeptide backbone around the axis of the rod.

α helical coiled coil: An element of the three-dimensional structure of proteins that consists of two α helical rods helically wound around one another.

triple helix: An element of the three-dimensional structure of proteins that consists of a helical arrangement of three strands of a polypeptide backbone around an axis.

Representing Protein Sequence and Three-Dimensional Structure in Two Dimensions

Rosemarie Swanson

The 3-D (physical) structure of a protein is complicated and hard to remember in detail, but it is coarsely described by a ribbon or rope making a tortuous path through space. The "ribbon" describes the path in space of a linear polymer of amino acids. Thus, an alternative linear representation of the protein is simply the "text" giving its amino acid sequence. The two descriptions give the viewer very different and independent pieces of information, and one cannot be inferred from the other. Both kinds of information are needed in a correlated form. A diagram is presented which combines the two kinds of information in a schematic 2-D (plane) projection of the 3-D structure with the amino acid sequence written along the trace of the projection. Such a diagram provides a compact, easy-to-use summary of the approximate relative positions of the amino acids in space. Knowledge of the space positions of amino acids in a protein is important for understanding its biochemical function, for modifying it by genetic engineering, and for developing an understanding of how a 1-D amino acid sequence dictates the 3-D structure of a protein.

The twenty different amino acids specified by the genetic code share an aspect with the letters of the English alphabet. Like the letters of the alphabet, the amino acids code information by their linear sequence. That is, the linear sequence of amino acids in a protein contains the information that causes the polymer to contort into the structure of the enzyme or protein. A naturally occurring linear sequence of amino acids is a biochemical "sentence" with a "meaning": Its "meaning" is the structure of a particular enzyme or protein. Different sequences form different structures, generally speaking, although amino acids may be substituted by similar ones[1] without changing the structure.

The structure of a protein — the path in space that its linear amino acid chain assumes and the way it folds up — determines its function. The importance of the space path (*fold*) for protein function is that it puts amino acids in position to catalyze a chemical reaction, to hold reactants in place, to sequester metal ions, or otherwise to contribute favorably to the action of the protein. Predicting the space path on the basis of an amino acid sequence alone is an unsolved problem of great interest to biochemists. (Some important partial solutions[2] have, however, been devised recently.)

Although space paths cannot be predicted in general, patterns of folding do exist. Two particularly important patterns are *helices* and *sheets*.[3] These patterns are known as *secondary structure*. Secondary structure is dictated by the unique side chain portions that distinguish the amino acids from each other, but it is mediated by the uniform main chain portions that allow the amino acids to polymerize.

A *helix* in a protein is just what the name suggests: A spiral formed by a section of the polymer chain. (Figure 1* shows an atomic model of a protein domain which includes two helical sections of chain. Figure 2(a)† shows a schematic redrawing of Fig. 1. The helical portions are the sections marked 25 to 33 and 69 to 77. The corresponding portions of Fig. 1 can be identified by comparison). The uniform main chain portion forms a spiral backbone and the non-uniform, information-carrying side chains project out from the "surface" of the spiral. The spiral structure is reinforced by weak cross-links (hydrogen bonds) between adjacent turns of the helix. The cross-links exist because each backbone link has both a donor site and an acceptor site for a hydrogen bond. These sites project in opposite directions and are aimed roughly at right angles to the line of the backbone link.

---

*Figure 1 was produced using atom coordinates for the enzyme ATCase and the program FRODO running on a Digital Equipment Corp. VAX connected to an Evans and Sutherland PS330 graphics machine. Atom coordinates for hundreds of proteins are available from the Brookhaven Protein Data Bank (Web address: http://pdb.pdb.bnl.gov/). FRODO is one of several different interactive molecular graphics programs that run on a variety of computers. Web source of information on molecular modeling software: http://www.nih.gov/molecular_modeling/guide_documents/.

†Figure 2 was drawn on the basis of the ATCase atom coordinates displayed with FRODO, using the program MacDraw II, available from Claris Corp., Mountain View, CA 94043, USA.

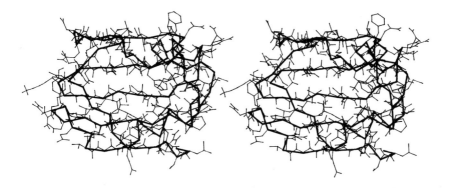

Fig. 1. Stereo view of an atomic model of a protein domain. (Inexpensive stereo viewers may be purchased from reference 5). The heavy line emphasizes the path of the polymer chain. The model shown is the allosteric domain of the regulatory chain of aspartate carbamoyltransferase. The domain contains 100 amino acids (only 93 can be located by x-ray crystallography) and has the function of binding nucleotides (see *pyrimidine nucleotides* in glossary), which regulate the activity of the enzyme in accord with the metabolic requirements of the *E. coli* cell for pyrimidine nucleotides.

The hydrogen bonding sites on the backbone links permit the formation of another kind of regular hydrogen bond cross-linked structure, the beta *sheet.* A sheet in a protein is also like what the name suggests: A planar structure with two surfaces. A sheet is formed from a set of adjacent parallel (or antiparallel) stretches (*strands*) of backbone links. (Figure 2(a) indicates the strands of the beta sheet with five long straight arrows; the arrow marked 41 to 46 is an example. The corresponding regions of Fig. 1 can again be located by comparison.) The amino acid side chains along each strand of the sheet project alternately above and below the surface of the sheet. The strands are cross-linked by hydrogen bonds in the plane of the sheet.

Despite these various regularities, the visual representation of all the atoms of a protein structure yields a very complex image, as demonstrated in Fig. 1. The information density is so high that the path of the chain is hard to follow and it is difficult to locate individual amino acids of interest, even though the structure[4] shown is relatively simple. A greatly abstracted representation that retains much of the spatial information yet makes the chain path unmistakable and the amino acid sequence literally readable is shown in Fig. 2(b) in the same orientation as the image in Fig. 1.

In Fig. 2(b), the five arrow-like boxes enclose five stretches of amino acids (five beta sheet strands) which align themselves in an antiparallel orientation

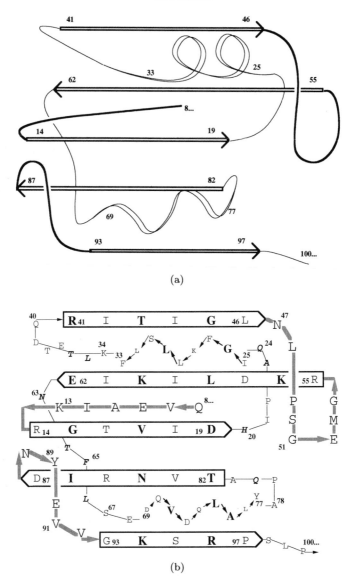

(a)

(b)

Fig. 2. (a) Simplified drawing of the same chain path as in Fig. 1. The numbers are the sequence numbers of the amino acids in the corresponding chain positions. (b) Schematic representation of the same structure as in Fig. 1, showing the amino acid types in their appropriate spatial relationships but representing them as letters instead of as chemical structures. See text for additional explanation.

and form cross-linking hydrogen bonds to make a relatively rigid internal framework for the protein structure. As mentioned above, the side chains of the amino acids project above or below the plane formed by the five strands, alternating up and down along the strand. "Up" (toward the viewer, above the plane of the paper) is indicated by **boldface**, "down" (away from the viewer, below the plane of the paper) by plain text. The "up" side chains are in register from strand to strand across the sheet, although the strands themselves need not be aligned with respect to their beginnings and ends. Above and below the plane of the sheet are parts of the polymer chain that connect the strands and contact other molecular units. Parts of the chain that are above the sheet are connected by thick gray lines, and parts that lie below the sheet are connected by thin black lines. Two helices lie below the sheet. The amino acid side chains project out from the surface of the cylinder that the chain path circumscribes in the regions where it forms helical turns. Amino acids whose side chains project toward the viewer from a helix are in **bold**, and side chains that project from the surface of the helix that is away from the viewer are shown in a smaller type size. Arrows indicate the direction of the polymer chain in the helices and the *loops* connecting the beta sheet strands. The schematic diagram presented here summarizes considerable structural and sequence information in a well-organized and easily retrievable form.

Acknowledgements

This work has been supported by the National Institute of General Medical Sciences (Research Grant GM-33191) and the Robert A. Welch Foundation (A-915).

References

1. Swanson, R., "A unifying concept for the amino acid code", *Bull. Math. Biol.* **46**, 187–203 (1984).
2. Bowie, J.U., Luthy, R. and Eisenberg, D., "A method to identify protein sequences that fold into a known three-dimensional structure", *Science* **253**, 164–170 (1991).
3. Dickerson, R.E. and Geis, I., *The Structure and Action of Proteins* (Benjamin Cummings Publishing Company, Inc., 1969).
4. Gouaux, J.E., Stevens, R.C. and Lipscomb, W.N., "Crystal structure of aspartate carbamoyltransferase ligated with phosphonacetamide, malonate and CTP

or ATP at 2.8 Anstrom resolution and neutral pH", *Biochemistry* **29**, 7702–7715 (1990).

5. Stereopticon 707, Taylor Merchant Corp., New York, NY 10001, USA, Tel: (212) 757–7700.

6. Branden, C. and Tooze, J., *Introduction to Protein Structure* (Garland Publishing, Inc., 1991).

7. Swanson, R., "A unifying concept for the amino acid code", *Bull. Math. Biol.* **46**, 187–203 (1984).

Glossary

aspartate carbamoyltransferase (ATCase): An enzyme engaged in an early step in the production of cytosine ("C") and uracil ("U"), two of the four "letters" in the genetic code. The enzyme is closely controlled in the cell and slows down when levels of the end products, CTP and UTP, are high enough. How this regulation is achieved is not well-understood at the molecular level. See also *pyrimidine nucleotides*.

pyrimidine nucleotides: The smaller of the two major types of nucleic acid units found in genetic materials. Purines are the other major type. In nucleic acid double helices, pyrimidines pair with purines (C with G, T or U with A). CTP, UTP, and TTP are abbreviations for the genetically significant pyrmidine nucleotides, representing cytidine triphosphate, uridine triphosphate and thymidine triphosphate, respectively. See also *purine nucleotides*.

purine nucleotides: Adenosine triphosphate (ATP) and guanidine triphosphate (GTP) are the genetically significant purine nucleotides. These forms contain the sugar ribose and are found in RNA. d-ATP and d-GTP refer to the forms containing 2'-deoxy ribose and are found in DNA. See also *pyrimidine nucleotides*.

Visual Display of Sequence Conservation as an Aid to Taxonomic Classification Using PCR Amplification

Peter K. Rogan, Joseph J. Salvo, R. Michael Stephens
and
Thomas D. Schneider

By comparing corresponding gene sequences from three or more organisms, we can deduce relationships which can be used for biological classification. Portions of a gene with the greatest number of differences (polymorphisms) generally produce the largest amount of useful data for classification of new specimens. In order to identify highly polymorphic regions, we displayed the 28S ribosomal RNA gene as a *sequence logo*.[15] In this representation, the height of letters measures the degree of sequence conservation among several species. We then picked a region which has two conserved motifs surrounding a highly variable domain. The conserved portions are the same in many organisms, so we were able to use the polymerase chain reaction (PCR) technique to amplify the variable portion in the middle. When sequenced and compared, these amplified pieces of DNA will allow taxonomic classifications to be made.

Introduction

Taxonomic classification of organisms requires the study of both the similarities and differences between the organisms. Genetic homology and mutations are analogous to the morphological criteria that have classically been used to deduce evolutionary relationships. Because they are so easy to obtain, nucleic-acid sequences of equivalent genes in different species are now used to determine taxonomic relationships between the species.[20]

If two different species have identical genes, then we have little to say about how they evolved from a common ancestor. When two genes differ at one or more sites, the number of nucleotide changes indicates how much the two organisms have diverged from a common ancestor. This paper describes a method for identifying and then amplifying regions of sequence divergence.

First, regions of DNA that have similar features are located and used to align the sequences. DNA sequence regions that are similar in more than one species are said to be "conserved". This conservation can be visualized with a sequence logo,[15] which displays the nucleotides that are present in multiple aligned sequences. These logos can be used to locate a region of sequence divergence surrounded by two regions of conservation. DNA primers are designed to anneal to the conserved regions for polymerase chain reaction (PCR) amplification.[12] These primers amplify both conserved regions and the divergent DNA sequences between them. The variable region between the two primer sequences can be used to construct dendrograms which depict the taxonomic relationships between different organisms. This method simultaneously satisfies the requirement of PCR for two conserved DNA sequences from which to perform DNA amplification, and the requirement of taxonomy for a highly variable DNA region from which to construct evolutionary trees.

The sequences of those RNA molecules, which are integral components of the ribosome, can be used to identify functional genetic differences between species. All organisms employ ribosomes to carry out the important biological task of protein synthesis, so ribosomal subunits have been structurally and functionally conserved throughout the eons. The sequences of ribosomal RNAs from widely differing species can be aligned and then the differences between these sequences specify the evolutionary or phylogenetic relationships between the organisms.[20]

Methods and Results

Eukaryotic ribosomes contain several conserved RNAs, the largest of which is called the 28S ribosomal RNA (rRNA). In order not to bias the analysis to regions studied previously,[7,20] we chose to use full length 28S sequences in the sequence alignment. Sequences having the broadest possible taxonomic distribution were selected in order to maximize species diversity (sequences of plant, animal, fungal, and protistan origin were used). The corresponding 28S ribosomal DNA (rDNA) sequences were obtained from the GenBank genetic sequence repository.[1] Sequences were aligned by a "rectification algorithm"[3,4] in which the human sequence was chosen to be a reference sequence. The human sequence was aligned in a pairwise fashion with each of the other sequences, and then the modified reference sequence (containing gaps) was realigned with the other sequences again to produce the final alignment. Although other alignment strategies might generate other solutions, they were not explored because our studies produced satisfactory results (see below).

A sequence logo was created from the aligned 28S rDNA sequences, and the region shown in Figs. 1 and 2 was one identified as having two conserved regions surrounding a divergent region. The horizontal axis represents nucleotide positions along the DNA whereas the vertical axis measures the degree of conservation at the same position in the various species. The vertical scale is given in bits of information, which measures the number of choices between two equally likely possibilities. The choice of one base from the four possible bases requires two bits of information.

The two bits correspond to two choices. For example, the first choice could determine whether the base is a purine or a pyrimidine and the second choice would specify which purine or pyrimidine is present. Thus, at position 100 of Fig. 1, all of the 28S sequences have a C so that the position has two bits of conservation. In the logo (Fig. 2), a C appears at position 100 with a height of (almost) 2 bits. (A small sample correction prevents it from being exactly 2 bits high.[16])

For those positions where two equally likely bases occur, there is only one bit of information. This is because a choice of two things from four is equivalent to a choice of one thing from two. Position 86 is an example of this, in which five of the sequences contain A and four have T in Fig. 1. This position is therefore about one bit high in Fig. 2. The relative frequency of the bases determines the relative heights of the letters, and since A is more frequent, it is placed on top. Positions in which all four bases are equally likely not conserved and so have zero heights on the logo. When the frequencies of the bases are other than 0, 50, or 100%, the heights still measure the conservation at each position, but the calculation is a bit more complicated.[9,13,14,16] However, this method permits comparisons of the height of one position with any other.[17,18]

The sequence logo was used to choose the two PCR primers, as shown in Fig. 2, according to the following three criteria:

P.K. Rogan et al.

```
                10        20        30        40
   Hum  -----GTCCG GTGAGCTCTC GCTGGCCCTT GAAAATCCGG GGGAG
   Lem  --GGTGTCCG GTGCGCCCCC GGCGGCCCTT GAAAATCCGG AGGAC
 Mrace  ------TCTG GTGCATTCAC AACGATCCTT GAAAATCCAA GGGAA
  Rice  --GGTGTCCG GTGCGCCCCC GGCGGCCCTT GAAAATCCGG AGGA-
 Slime  --------GC GGTCTCCTTC CGTTGCCCTA GAAAAGCTGG CAGAT
   Tom  --GGTGTCCG GTGCGCTCCC GGCGGCCCTT GAAAATTCCG GAGGA
  Worm  GTGGTGTCTC GTGCTCTTTG AACGGCCCTT AAAACACCAA GGGAG
  Xlrn  -GGCGTCCGG TGAGCTTCTC GCTGGCCCTT GAAAATCCGG GGGAG
 Yeast  --TGGCTCCG GTGCGCTTGT GACGGCCCGT GAAAATCCAC AGGA-

                50        60        70        80        90
   Hum  AGG-- ----GTGTAA ATCTC-GCGC CGGGCCGTAC CCATATCCGC
   Lem  CG--- ----AGTG-- CCGCCCGCGC CCGGTCGTAC TCATAACCGC
 Mrace  A---- -----GAATA ATTTTCTCGC CTAGTCGTAC TCATAACCGC
  Rice  ----- ---CCGAGTA CCGTCCACGC CCGGTCGTAC TCATAACCGC
 Slime  GGGTG AAACGTGTTG TCCTTCG-GT TGAACCGTAC CTA-ATCCGC
   Tom  CCGAA TGCCGT---- ---TCCACGC CCGGTCGTAC TCATAACCGC
  Worm  GCTAT -------TAA TT---TGCAC TCAATCGTAC CGATATCCGC
  Xlrn  AGG-- ----GTGTAA ATCTCTGCGC CGGGCCGTAC CCATATCCGC
 Yeast  ----- ---AGGAATA GTTTTCATGC TAGGTCGTAC TGATAACCGC

                100       110       120       130
   Hum  AGCAGGTCTC CAAGGTGAAC AGCCTCTGGC ATGTTGGAAC AATGT
   Lem  ATCAGGTCTC CAAGGTGAAC AGCCTCTGG- TCGATGGAAC AATGT
 Mrace  AGCAGGTCTC CAAGGTGAAA AGCCTCTAG- TTGATAGAAC AATGT
  Rice  ATCAGGTCTC CAAGGTGAAC GACCTCTGGC -CAATGGAAG AATGT
 Slime  AGCAGGTCTC CAAGATGAGC AGTCTCTGGC GCATAGAACA AAGTA
   Tom  ATCAGGTCTC CAAGGTGAAC AGCCTCTGG- TCGATGGAAC AATGT
  Worm  ATTAGGTCTC CAAGGTGAAC AGCCTCTAG- TCGATAGAAT AATGT
  Xlrn  AGCAGGTCTC CAAGGTGAAC AGCCTCTGGC ATGTTAGAAC AATGT
 Yeast  AGCAGGTCTC CAAGGTGAAC AGCCTCTAG- TTGATAGAAT AATGT

                140       150       160       170
   Hum  AGGTA AGGGAAGTCG GCAAGCCGGA TCCGTAACTT CGG
   Lem  AGGCA AGGGAAGTCG GCAAAATGGA TCCGTAACTT CGG
 Mrace  AGATA AGGGAAGTCG GCAAAATAGA TCCGTAACTT CGG
  Rice  AGGCA AGGGAAGTCG GCAAAACGGA TCCGTAACTT CGG
 Slime  GCGTA AGGGAATTCG GCAAGCCGGA TTCGTAACTT CGG
   Tom  AGGCA AGGGAAGTCG GCAAAATGGA TCCGTAACTT CGG
  Worm  AGGTA AGGGAAGTCG GCAAACTAGA TCCGTAACTT CGG
  Xlrn  AGGTA AGGGAAGTCG GCAAGTCAGA TCCGTAACTT CGG
 Yeast  AGATA AGGGAAGTCG GCAAAATAGA TCCGTAACTT CGG
```

Fig. 1. Aligned sequences for part of the 28S rDNA from nine species. Hum: *Homo sapiens* (humans); Lem: *citrus limon* (lemon); Mrace: *Mucor recemosis* (zygomycete, a fungus); Rice: *Oryza sativa* (rice); Slime: *Physarum polycephalum* (slime mold); Tom: *Lycopersican esculentum* (tomato); Worm: *Caenorhabditis elegans* (nematode worm); Xlrn: *Xenopus laevis* (South American toad), and Yeast: *Saccharomyces cerevisiae* (baker's yeast).

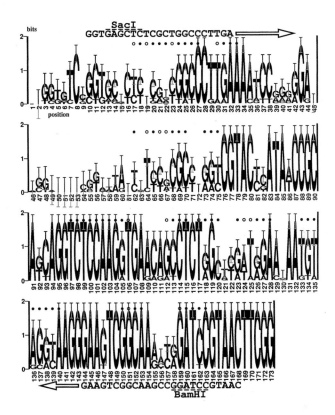

Fig. 2. Sequence logo created from the sequences shown in Fig. 1. The horizontal axis represents the position of nucleotides in the alignment; these correspond to Fig. 1. The vertical axis represents information (i.e., sequence conservation) in bits. Error bars indicate the standard deviation of the height of each stack. (This represents the expected variation of the conservation due to the small (finite) sample of aligned sequences.) See text for further description. The sequence of the *Sac*I primer is as shown. The complement of the *Bam*HI primer is shown for clarity. (The actual sequence of the *Bam*HI primer on the other strand of DNA is found by switching A's with T's and C's with G's and writing the sequence backwards: 5′ GTTACGGATCCGGCTTGCCGACTTC 3′.) The primers are identical to the human 28S rDNA sequences. The 5′ and 3′ terminal coordinates of the PCR product correspond to positions 2698 and 2849 of the human sequence (GenBank entry HUMRGM, accession number M11167). Proposed RNA structure in *X. laevis* 28S RNA[2] is shown by paired bases (●) and G-U base pairs (○). Unmarked bases are single stranded. The sequence logo programs are written in standard (i.e., portable) Pascal,[6] and most have been translated into C. They are available by anonymous ftp (file transfer protocol) from ncifcrf.gov in directory pub/delila. See the README file there for further information. Electronic mail contact: toms@ncifcrf.gov.

- They are in regions of high conservation, and surround regions of low conservation.
- The 3′ termini cover regions that are not variable, so that the primer end which is extended by the DNA polymerase is always properly annealed to the DNA.
- The oligonucleotide primers are not self-complementary and do not base pair to each other.

Because these primers were also designed to guarantee amplification of the human sequence, the 5′ terminus of the left primer was not highly conserved. This had no effect on cross species amplification. The primers contain restriction sites useful for subsequent cloning of the amplification products (results not shown). The primers cover DNA from base 10 through 168 (Fig. 1), so the amplified human product was predicted to be 159 base pairs long. When PCR reactions were carried out on genomic DNA purified from 12 different organisms a major product was observed (Fig. 3). Except for *H. sapiens*, none of the species tested were included in the original sequence alignment. This result demonstrates that the method can be applied to new species. A number of mammalian species were amplified as well as several different classes of fungi. The set of fungal species that were successfully tested is particularly significant, as ribosomal sequences within this subkingdom exhibit a very high degree of diversity.[19] The DNA sequences of the cloned *H. sapiens*, *M. musculus*, and *S. cerevisiae* amplification products corresponded precisely with published reports (data not shown), suggesting that amplification was of high fidelity and that contaminating genomic templates were not present.[10] The phylogenetic relationships inferred from the 28S sequences of the other species were generally compatible with the known taxonomic relationships between these organisms.[10]

Minor amplification products are observed in several of these reactions (*L. catta*, *P. infestans*, *M. musculus*, *P. pinus*, and *S. cerevisiae* lanes). The fragments may have arisen from amplification at sites within these genomes which happen to be complementary to our primers. This possibility is, in part, a function of the annealing temperature of the PCR reaction, which was deliberately chosen to be permissive for amplification of rDNA from a wide variety of extant species. Alternatively, since organisms have many copies of the 28S gene, some of these may have large variations in the amplified regions. In particular, the region selected for amplification is adjacent to a target sequence which is commonly interrupted by a site-specific transposable element

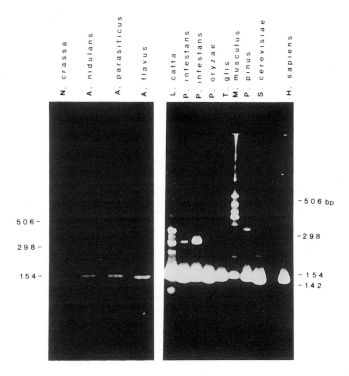

Fig. 3. PCR amplification reaction products. The following conditions were used for PCR amplification: 1 unit of Taq DNA polymerase (Perkin Elmer Cetus) was used. After incubation of oligonucleotide primers and genomic DNA for 4′ at 94°C, 40 cycles were performed at 94°C, 1′; 55°C, 1′; 72°C, 2′ followed by a synthesis step (72°C,7′). The products were separated on a 4% Nusieve agarose gel (Seakem). The gel was stained with ethidium bromide and photographed under ultraviolet light. Reactions (from left to right) are: *Neurospora crassa*, *Aspergillus nidulans*, *Aspergillus parasiticus*, *Aspergillus flavus* (fungal species), *Lemur catta* (brown lemur), *Phytophthora infestans* (Irish potato blight of 1840, strain 506), *Phytophthora infestans* (strain 10126), *Pyricularia oryzae* (rice blast fungus), *Tupaia glis* (tree shrew), *Mus musculus* (domestic mouse), *Pichia pinus* (a soil fungus), *Saccharomyces cerevisiae* (baker's yeast), and *Homo sapiens*.

in some organisms (i.e., insects.[5]) Thus, the evolutionary instability of this region of the 28S gene may give rise to variants which, upon amplification, could produce these minor species-specific PCR products. These artifacts do not interfere with subsequent isolation, cloning, and sequencing of the desired products.[11]

Some nucleotide sequences within the RNA molecule are complementary to one another because the formation of these duplexes is required for normal ribosome function. The proposed secondary structure of base-paired nucleotides in the amplified region of the 28S RNA molecule[2] does not correlate with the information content at each position (Fig. 2). In part, this is because the information measure we used shows conservation at each position and ignores possible cross correlations. For example, if an A is paired to U somewhere in the structure, and then the A is mutated to C during evolution, the other base must become G for the structure to be maintained. However, compensatory mutations are not always found in duplex structures.[8] The fixed bases would display high information content in a sequence logo.

In the 28S RNA sequence that we analyzed, the predicted hairpin structures do not always exhibit a high information content (e.g., positions 62-74), although at least one of the potential duplexes is highly conserved (e.g., positions 97-106). Conversely, although the most variable stretch of nucleotides in this region is found within a predicted single stranded region (positions 33-61), this is not always the case, as some single stranded domains are also highly conserved (e.g., positions 76-96, 164-173). This is consistent with the possibility that these sequences make specific interactions with ribosomal proteins. It is not surprising then, that the 28S sequence logo derived here does not uniformly correlate with the postulated secondary structure.

Discussion

The consensus sequence of an aligned set of sequences is constructed by choosing the most frequent base at each position. Consensus sequences were not used to locate conserved and non-conserved regions because they are not quantitative and they destroy data about the frequency of base occurrences. Because the sequence logo measures sequence conservation in a quantitative way, variable regions that may be useful in studying phylogeny can be rationally selected. Although we did not do so, it may be possible to design degenerate PCR primers where no single nucleotide was present in all of the aligned sequences. Such an oligonucleotide should be complementary to most of the targets, but unfortunately it would also be complementary to many other sequences.

In summary, the sequence logo reveals evolutionary and, by inference, functional conservation of specific nucleotide sites in the ribosomal sequence. However, when applied in molecular phylogenetic analysis, it is simply a tool for assessing the most variable or conserved segments in the rDNA sequence. A highly variable domain flanked by highly conserved segments was identified in the sequence logo and the two conserved elements were used to design oligonucleotide primers for PCR amplification. These primers were shown to be suitable for amplification of the variable region in a wide variety of eukaryotes. The DNA sequence of these PCR amplification products can be used to determine taxonomic relationships or to quickly place a previously unidentified species on the evolutionary tree.

References

1. Burks, C., Cassidy, M., Cinkosky, M.J., Cumella, K.E., Gilna, P., Hayden, J.E.-H., Keen, G.M., Kelley, T.A., Kelly, M., Kristofferson, D. and Ryals, J., "GenBank", *Nucl. Acids Res.* **19**, 2221–2225 (1991).
2. Clark, C.G., Tague, T.W., Ware, V.C. and Gerbi, S.A., "*Xenopus laevis* 28S ribosomal RNA: Secondary structure model and its evolutionary and functional implications", *Nucl. Acids Res.* **12**, 6197–6220 (1984).
3. Feng, D.F. and Doolittle, R.F., "Progressive sequence alignment as a prerequisite to correct phylogenetic trees", *J. Mol. Evol.* **25**, 351–360 (1987).
4. Higgins, D.G. and Sharp, P.M., "Fast and sensitive multiple sequence alignments on a microcomputer", *CABIOS* **5**, 151–153 (1989).
5. Jakubczak, J.L., Burke, W.D. and Eickbush, T.H., "Retrotransposable elements R1 and R2 interrupt the rRNA genes of most insects", *Proc. Natl. Acad. Sci. USA* **88**, 3295–3299 (1991).
6. Jensen, K. and Wirth, N., *Pascal User Manual and Report* (Springer-Verlag, 1975).
7. Lane, D.J., Pace, B., Olsen, G.J., Stahl, D.A., Sogin, M.L. and Pace, N.R., "Rapid determination of 16S ribosomal RNA sequences for phylogenetic analyses", *Proc. Natl. Acad. Sci. USA* **82**, 6955–6959 (1985).
8. Michel, F. and Westhof, E., "Modelling of the three-dimensional architecture of group I catalytic introns based on comparative sequence analysis", *J. Mol. Biol.* **216**, 585–610 (1990).
9. Pierce, J.R., *An Introduction to Information Theory: Symbols, Signals and Noise*, 2nd ed. (Dover Publications, 1980).
10. Rogan, P.K., Salvo, J.J. and Tooley, P.W., "Amplification of 28S ribosomal DNA with universal PCR primers", *Proceedings of the IV International Congress of Systematic and Evolutionary Biology*, University of Maryland, College Park, MD, Vol. 2 (1990) pp. 393.
11. Rogan, P.K. and Salvo, J.J., "High-fidelity amplification of ribosomal gene sequences from South American mummies", *Ancient DNA*, eds. B. Herrmann and S. Hummel (Springer-Verlag, 1994).

12. Saiki, R.K., Scharf, S., Faloona, F., Mullis, K.B., Horn, G.T., Erlich, H.A. and Arnheim, N., "Enzymatic amplification of β-globin genomic sequences and restriction site analysis for diagnosis of sickle cell anemia", *Science* **230**, 1350–1354 (1985).
13. Schneider, T.D., "Theory of molecular machines. I. Channel capacity of molecular machines", *J. Theor. Biol.* **148**(1), 83–123 (1991a). (Note: The figures were printed out of order. Figure 1 is on pp. 97).
14. Schneider, T.D., "Theory of molecular machines. II. Energy dissipation from molecular machines", *J. Theor. Biol.* **148**(1), 125–137 (1991b).
15. Schneider, T.D. and Stephens, R.M., "Sequence logos: A new way to display consensus sequences", *Nucl. Acids Res.* **18**, 6097–6100 (1990).
16. Schneider, T.D., Stormo, G.D., Gold, L. and Ehrenfeucht, A., "Information content of binding sites on nucleotide sequences", *J. Mol. Biol.* **188**, 415–431 (1986).
17. Shannon, C.E., "A mathematical theory of communication", *Bell System Tech. J.* **27**, 379–423, 623–656 (1948).
18. Shannon, C.E. and Weaver, W., *The Mathematical Theory of Communication* (University of Illinois Press, 1949).
19. Walker, W.F., "5S and 5.8S ribosomal RNA sequences and protist phylogenetics", *Biosystems* **18**, 269–278 (1985).
20. Woese, C.R., "Bacterial evolution", *Microbiol. Rev.* **51**, 221–271 (1987).

Glossary

5′ and 3′: The ends of a single strand of DNA are named 5′ and 3′ after the names of carbon atoms in the deoxyribose sugar. These define direction along the DNA. In living organisms, the strands of DNA are always synthesized from the 5′ end toward the 3′ end by DNA polymerases. Ends are also called "termini".

complement: In double stranded DNA or RNA, the bases on opposite strands are said to be "complementary" because their surfaces fit together nicely. A is complementary to T (or U) and G is complementary to C.

DNA: Deoxyribonucleic acid, the genetic material. There are four chemical letters (called bases) in DNA: A, C, G, and T. These are connected together by alternating deoxyribose sugar and phosphate groups to form long molecules. A gene is composed of a series of these bases. The order of bases in a DNA is called its "sequence", and "sequencing DNA" means to read the DNA letters. Usually, two strands of DNA are wound around each other to form the famous "double helix". It looks like a twisted ladder, with the rungs made of two bases called "base pairs". If there is an A on one side of a rung, there will be a T on the other (and *vice versa*). Also, if there is a C on one side of a rung, there will be a G on the other (and *vice versa*). During DNA replication, each strand is copied by an enzyme called DNA polymerase. This produces two "daughter" DNA molecules identical to the original parent molecule.

enzyme: A protein which catalyzes (speeds up) a chemical reaction. An example is DNA polymerase, which creates DNA strands by polymerizing the four bases together.

eukaryote: (literally, true nucleus) A living organism whose cells contain membrane bounded nuclei in which their DNA is held. Examples are humans and plants. Procaryotes (literally, before nuclei) do not have nuclei. Examples are bacteria and their viruses.

gene: A segment of DNA which codes for an RNA or protein. Genes perform distinct functions in the cell.

GenBank: One of the three world repositories for genetic sequence information. (See Burks, C. *et al.*, *GenBank. Nucl. Acids Res.* **19**, 2221–2225 (1991).)

interstitial region: In the context of amplifying a piece of DNA by PCR, the region between two PCR primers.

PCR: See polymerase chain reaction.

polymerase chain reaction: (PCR) A method of making many copies of ("amplifying") a DNA. PCR can be started even from a single molecule. The method relies on the ability of the enzyme DNA polymerase to make copies of DNA in the test tube. To start a PCR reaction, one puts some DNA polymerase into a tube, along with the four bases, some DNA and two DNA primers. Primers are pieces of synthetic DNA 15 to 20 bases long. The solution is heated to make the DNA open up. This allows the primers to stick to the DNA when the solution is cooled. They stick in places where their bases match (complement) the bases of the DNA (A to T and C to G). The DNA polymerase binds to the primer/DNA complex and starts replicating the DNA. The two primers are chosen so that they are not too far apart, and so that the DNA polymerases which start the replications will be going toward each other. Because they are on different strands, they will not collide, but will pass each other by. This produces two molecules identical to the parent molecule, but starting at the primers. The solution is heated up again to separate all the strands, and when it is cooled, the primers stick to the DNA again. After a series of heating and cooling cycles, the region of DNA between (and including) the primers is replicated far faster than any other piece of DNA. It is therefore "amplified". The cyclic heating and cooling is performed by a machine, and special heat insensitive DNA polymerases (from hot springs bacteria) are often used.

protein: A long string of amino acids, which folds up into a particular shape. Proteins can be used for structural purposes (e.g., keratin in hair, nails and horns) or they can do things (e.g., cenzymes like DNA polymerase).

ribosome: The large enzyme in cells which translates RNA into protein. The ribosome is made up of many pieces of RNA and protein.

RNA: Ribonucleic acid, a copy of the genetic material (see DNA). In RNA there are four chemical letters (called bases): A, C, G and U (instead of T). Also, the connecting backbone has ribose instead of deoxyribose. RNA is a copy of the DNA which usually survives for a only few minutes in the cell before being broken down. Ribosomal RNAs, however, last indefinitely. The DNA copy of the ribosomal RNA is called an rDNA. RNA usually has only one single strand, but small portions of it can loop back on themselves to make a so-called "hairpin". These twist together like the structure in DNA. Hairpin loops only form if sequences of at least four base pairs in length are complementary. RNA molecules fold up into complex three-dimensional structures. We usually do not know what these "tertiary" structures look like in detail, but maps showing which parts contact which other parts by complementary base pairing — called the "secondary structure" — can often be made. The "primary structure" is the sequence itself. Some rather complex secondary structures have been discovered in RNA, especially in the ribosomal RNAs.

taxonomy: The classification of organisms into related groups. The final product is an evolutionary tree. By counting the number of changes between sequences from two species, we can estimate how long it has been since they had a common ancestor.

Perceptible Features in Graphical Representations of Nucleic Acid Sequences

Jacques Ninio

and

Eduardo Mizraji

Computer calculations on sequences provide specific answers to specific questions. On the other hand, graphical representations of sequences are perhaps better suited for drawing one's attention towards unsuspected features. Three types of graphical coding of nucleic acid sequences, in which the four nucleotides G, C, A, T (or U) are replaced with graphical symbols, are presented. A vectorial representation, transforming the sequence into a planar trajectory, reveals the long-range statistical properties of the sequences. Intron-exon junctions or boundaries between secondary structure domains often coincide with changes in direction of the representative curve.

Massive DNA sequencing has produced to this date a number of striking results, among which:

(1) The genetic code is not universal;
(2) Genes are split in eukaryotes (and sometimes in other kingdoms) and reverse transcription plays a role in shaping them;
(3) Immunological diversity is largely combinatorial; and
(4) Some mitochondrial (or even nuclear) mRNAs are subject to "editing".

Most often, those who did the sequencing work were sufficiently astute to discover the revolutionary implication of their sequences, when there was one, without computer help.

The standard situation is less clear and straightforward. The biologist dealing with some authentic scientific problem reasons that a good way to learn about his system would be to clone the relevant genes and sequence them. Once this easily-funded project is undertaken, the biologist finds himself in possession of a collection of sequences — hundreds or thousands of A, T, G, C symbols in succession — from which nothing emerges. While writing a new grant application proposing the determination of more and more sequences in order to gain a finer understanding of reality, our biologist starts his quest for computer programs capable of handling the nucleotide strings and making a diagnosis; for example, "there are some knotty secondary structures close to the 3' end; there is a severe bias in codon usage; your sequence shows 100% homology with a previously published sequence."

Most often, such computer programs look for statistical constraints within or across sequences. All marks of "non-randomness" are highly cherished, and converted into "significant patterns". Yet, a substantial fraction of the "significant patterns" detected by the computer bear no relationship to the function under analysis.[1]

In order to deal with the sequence flood, it is perhaps important to develop more telling representations of the data. The computer would be discharged from most of the calculations, and the extraction of information would rely upon the cognitive capacities of the human brain. Several graphical or acoustic representations of sequences were proposed in a science-fiction spirit[2] in the past. However, this volume attests to the fact that, in this case, fiction was not so much in advance of reality. We summarize here our experience with previously published graphical representations.[3]

In the first type of coding, the sequence is simply printed on consecutive lines with regularly-spaced characters, the originality being in the design of the four characters replacing the usual symbols A, T, G, C (Fig. 1). This coding is illustrated with two sequences, the yeast iso-1 cytochrome c gene (top) and the origin of replication in S. typhimurium (bottom). Two classical dichotomies are embodied in this representation: Purine versus pyrimidine (up or down) and G or C versus A or T (full squares or pairs of bars).

This representation makes visible the polypurine and polypyrimidine stretches and the alternating purine-pyrimidine regions. For example, one

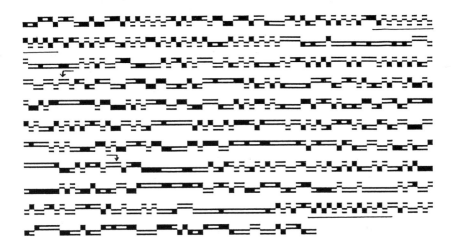

```
      ·⁻·▪  ATGC
      ▪⁻▫  GATC
```

Fig. 1. Four-symbol static coding. This representation is suitable for the detection of simple repetitive patterns, such as alternating purine-pyrimidine stretches. The sequence of the gene for yeast iso-1 cytochrome c[14] is represented at the top. The longest alternating purine-pyrimidine stretches are underlined. The arrows indicate the extremities of the coding part. The sequence at the bottom is that of the origin of replication in *Salmonella typhimurium*.[15] The GATC sequences are easily located but do not form conspicuous patterns. There are 80 characters in a row.

easily finds in the flanking regions before and after the coding sections of
the yeast iso-1 cytochrome c gene, two long alternating sections of 17 and
19 nucleotides. If one wishes, one can choose a different code for even and
odd nucleotides, and thus represent the alternating sequences as continuous
stretches. Then, however, their exact length becomes harder to appreciate.
Alternating purine-pyrimidine sequences are possible sites at which DNA may
adopt local conformations, and turn left instead of right. In the context of the
origin of life, Orgel[4] proposed that primitive genes were made of alternating
purine-pyrimidine sequences. Here, the sites of alternation lie outside the gene.

Bacterial origins of replication are rich in the tetranucleotide GATC. These
sequences can be spotted in Fig. 1 (bottom) but they do not catch the eye im-
mediately. Yet, the pattern is a most simple one. In general, the representation
of Fig. 1 is of little value for organizing the sequence into conspicuous patterns
other than the simple repetitive ones. This has two explanations.

First, the brain, unlike digital computers, deals better with an assembly of
complex symbols than with an equivalent collection of simple symbols. It is
easier for us to compare, say, 4283 and 3263 than their binary representations:
100001011011 and 11001011111. Second, the brain dislikes angulated shapes
and rather works on smooth curves. For a review of the principles of human
shape analysis, see reference 5. It would be possible to turn the crenels of Fig. 1
into sinusoids, but then the polypurine or polypyrimidine stretches would look
chaotic.

The representation of Fig. 2 embodies the same dichotomies as before. The
symbols are now curved ones. Purines have their concavity turned to the right
and pyrimidines to the left. The GC/AT dichotomy is coded by the size of
the characters, big for G or C, small for A or T. For each nucleotide, there
are three different symbols, the choice of which is context-dependent. When
there is a succession of symbols having their concavities turned to the same
side, all intermediate symbols are alleviated. One virtue of the system is its
legibility, even after severe reduction. A sequence remains interpretable at a
linear density of 10 nucleotides/cm.

The yeast iso-1 cytochrome c gene is thus represented in Fig. 2 (top). Large,
strictly complementary sequences form easily detectable symmetric patterns.
Imperfect complementarities may or may not be detected. The polypurine
or polypyrimidine stretches give obvious signals, but the alternating purine-
pyrimidine sequences have a rugged appearance and are not easily analyzed.

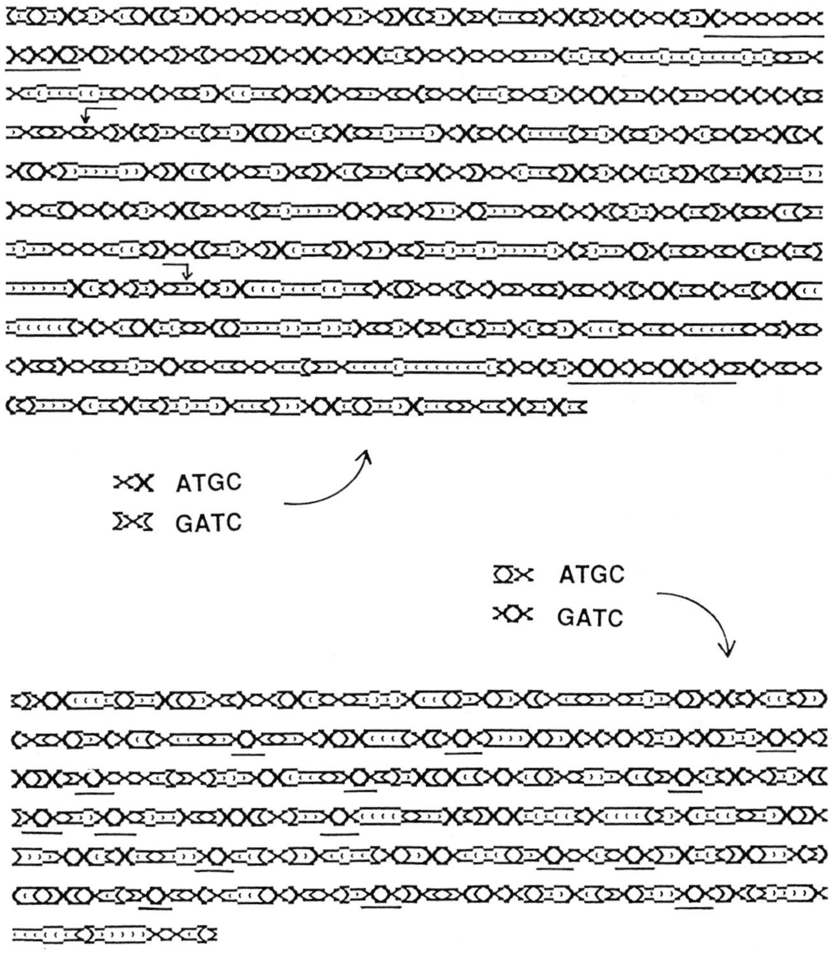

Fig. 2. Twelve-symbol static coding. The represented sequences are the same as in Fig. 1. This representation helps to organize the sequence into perceptually significant patterns. Long alternating purine/pyrimidine tracts are underlined. With the coding chosen for the top sequence, symmetric patterns correspond to strictly complementary sequences. With the coding shown at the bottom, GATCs are easily spotted.

The sequence appears organized into motifs. The origin of replication of
S. typhimurium is shown in Fig. 2 (bottom) using a different correspondence
between the nucleotides and their symbols. The GATC motifs then jump off
the page, together with a few tetranucleotides that differ from GATC by just
one base.

Our aim was to generate perceptually significant patterns so that salient
features could be at once recognized by the brain. It is amusing to note, in
this context, that perceptually poor graphical codings have been proposed[6] on
the grounds that they provide good reading material ... for the computer!

The vectorial representations, first documented by Hamori and Ruskin[7] and
by Swanson[8] are, we believe, the most promising ones, and several variants of
them have been designed.[3,7−11] In some variants, the DNA is represented by a
curve[7] or a solid[11] in 3-D space. Various aspects of the curve or the solid are
highlighted, according to the chosen viewpoint or the selected projection.

In our variant,[3] the representative curve is directly constructed in 2-D.
The four bases A, T (or U), G, C are represented by four vectors in 2-D space.

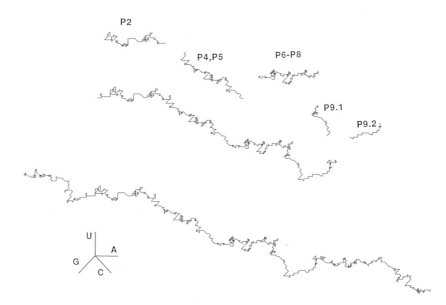

Fig. 3. Vectorial coding. Each nucleotide is represented by a vector in a plane. The sequence
shown in the middle is that of the self-splicing intron of *Tetrahymena pigmentosa*.[16] A larger
portion of the gene is shown below. Hairpins or hairpin clusters, labeled according to current
terminology[12] are shown separately at the top.

Current monitors produce decent straight lines along only four directions: The horizontal and vertical diections of the sheet and the $\pm45°$ directions. Accordingly, the directions of our four vectors (see Fig. 3) are defined by the cosine pairs $(1, 0)$, $(0, 1)$, $(s, -s)$, $(-s, -s)$, with $s = \sqrt{2}/2$. In this manner, the four bases are partitioned into two sets of two bases, the representative vectors of which are opposed two by two (the first of the above list with the third, the second with the fourth), yet do not cancel each other completely. This allows, in almost every case, the representative curve to be in a reasonably extended configuration. An analog to the viewpoint variability of the 3-D representations is introduced here by varying the assignments of particular vectors to particular bases.

A vectorial representation of the self-splicing intron of *Tetrahymena pigmentosa* is shown in Fig. 3. The drawing presents a succession of sections, each having a characteristic local direction. These sections correspond well with various domains of secondary structure of the molecule, according to the generally accepted model.[12]

The small subunit ribosomal RNAs of yeast and *E. coli* are shown in Fig. 4. The yeast 18S RNA seems to be structured in four domains, but the correspondence with the four domains of secondary structure[13] is rather loose. In the

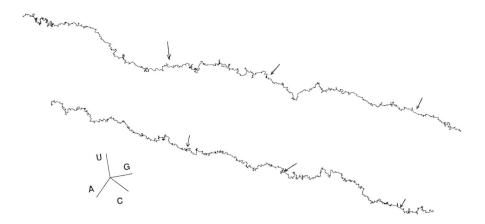

Fig. 4. Ribosomal RNA. The small subunit ribosomal RNAs from yeast[17] (top) and *E. coli*[18] (bottom) are represented vectorially, with the lengths of the vectors made inversely proportional to the frequencies of the bases they represent in each sequence. The arrows delineate the secondary structure domains, according to current models.[13]

same representation, the *E. coli* 16S RNA appears to be more homogeneous. It is not unlikely that in bacteria (or more generally, in organisms with small genomes), there are general selective constraints on DNA sequences which tend to promote statistical homogeneity all along the genes.

The last example is that of human α-hemoglobin (Fig. 5). The separation in introns and exons is obvious. The relationship between gene and pseudogene is less obvious. The positioning of the intron-exon junction suggests that the frontiers may have moved somewhat, due to nibbling of introns by adjoining exons. The introns of β-hemoglobin, shown in reference 3, seem unrelated to those of α-hemoglobin.

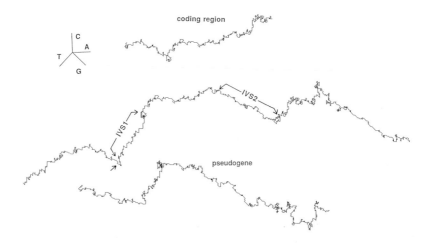

Fig. 5. Human α-hemoglobin. The sequence in the middle is that of a human α-hemoglobin gene.[19] The introns are labeled IVS1 and IVS2, and the complete coding region is shown on top. The bottom sequence is that of a pseudogene for human α-hemoglobin.[20]

One conspicuous property of vectorial representations is that they present compressed or expanded regions. Roughly, a compressed region is one in which there is a high turnover of the four nucleotides. In extended regions, there is a bias against one or more nucleotides. In all cases examined so far, the compressed/extended criterion failed to lead to biologically important insights. On the other hand, the dissection of the sequence in regions of constant local direction often reflects known biological features.

Three different correspondences between vectors and nucleotides were used for the three examples of Figs. 1–5. Twelve different codings are possible, plus their mirror-images. For each of the drawings presented here, the twelve independent codings were tried, and the most telling one was selected.

Each reader will have a slightly different reaction to the various drawings presented in Figs. 3–5. Once a feature is noticed, one can look for similar features, or radically different ones. While the computer gives unimaginative answers to the problem that was posed, the drawing helps to generate new questions and may suggest tentative answers. Having represented the sequences graphically, one can then switch back to the computer and perform the apppropriate statistical controls.

Acknowledgements

This work was supported by a grant from the defunct Agence de l'Informatique.

References

1. Ninio, J. and Mizraji, E., "String analysis and energy minimization in the partition of DNA sequences", *J. Mol. Biol.* **207**, 585–596 (1989).
2. Ninio, J., "L'explosion des séquences: Les années folles 1980–1990", *Biochem. Syst. and Ecol.* **11**, 305–313 (1983).
3. Mizraji, E. and Ninio, J., "Graphical coding of nucleic acid sequences", *Biochimie* **67**, 445–448 (1985).
4. Orgel, L.E., "Prebiotic polynucleotides and polypeptides", *Israel J. Chem.* **14**, 11–16 (1975).
5. Ninio, J., *L'empreinte des sens. Perception, mémoire, langage* (Editions du Seuil and Odile Jacob, 1991).
6. Lathe, R. and Findlay, R., "Machine-readable DNA sequences", *Nature* **311**, 610 (1984).
7. Hamori, E. and Ruskin, J., "H curves, a novel method of representation of nucleotide series, especially suited for long DNA sequences", *J. Biol. Chem.* **258**, 1318–1327 (1983).
8. Swanson, R., "A vector representation for amino acid sequences", *Bull. of Math. Biol.* **46**, 623–639 (1984).
9. Gates, M.A., "Simpler DNA sequence representations", *Nature* **316**, 219 (1985).
10. Pickover, C.A., "DNA vectorgrams: Representations of cancer genes as movements on a 2-D cellular lattice", *IBM J. Res. and Devel.* **31**, 111–119 (1987).
11. Pickover, C.A., "DNA and protein tetragrams: Biological sequences as tetrahedral movements", *J. of Molecular Graphics* **10**, 2–6 (1992).
12. Burke, J.M., Belfort, M., Cech, T.R., Davies, R.W., Schweyen, R.J., Shub, A., Szostak, J.W. and Tabak, H.F., "Structural conventions for group I introns", *Nucl. Acids Res.* **15**, 7217–7221 (1987).

13. Gutell, R.R., Weiser, B., Woese, C.R. and Noller, H.F., "Comparative anatomy of 16S- like ribosomal RNA", *Progress in Nucleic Acids Res. Mol. Biol.* **32**, 155–216 (1985).

14. Smith, M., Leung, D.W., Gillam, S., Astell, C.R., Montgomery, D.L. and Hall, B.D., "Sequence of the gene for iso-1-cytochrome c in *Saccharomyces cerevisiae*", *Cell* **16**, 753–761 (1979).

15. Zyskind, J.W. and Smith, D.W., "Nucleotide sequence of the *Salmonella typhimurium* origin of replication", *Proc. Natl. Acad. Sci. USA* **77**, 2460–2464 (1980).

16. Wild, M.A. and Sommer, R., "Sequence of a ribosomal RNA gene intron from *Tetrahymena*", *Nature* **283**, 693–694 (1980).

17. Rubtsov, P.M., Musakhanov, M.M., Zakharyev, V.M., Krayev, A.S., Skryabin, K.G. and Bayev, A.A., "The structure of the yeast ribosomal RNA genes. I. The complete nucleotide sequence of the 18S ribosomal RNA gene from *Saccharomyces cerevisiae*", *Nucleic Acids Res.* **8**, 5779–5794 (1980).

18. Carbon, C., Ehresmann, C., Ehresmann, B. and Ebel, J.-P., "The complete nucleotide sequence of the ribosomal 16-S RNA from *Escherichia coli*", *Eur. J. Biochem.* **100**, 399–410 (1979).

19. Liebhaber, S.A., Goossens, M.J. and Wai Kan, Y., "Cloning and complete nucleotide sequence of human 5'-α-globin gene", *Proc. Natl. Acad. Sci. USA* **77**, 7054–7058 (1980).

20. Proudfoot, N.J. and Maniatis, T., "The structure of a human α-globin pseudogene and its relationship to α-globin gene duplication", *Cell* **21**, 537–544 (1980).

Glossary

eukaryote: An organism whose cells contain their genetic material within nuclei.

purine: A class of two-ringed chemical components that includes the DNA bases adenine and guanine.

pyrimidine: A class of one-ringed chemical components that includes cytosine, thymine, and uracil.

exon: A coding sequence in the DNA of genes.

intron: An intervening sequence in the DNA of genes that is not part of the coding sequence.

pseudogene: A DNA sequence that is nearly homologous to a gene sequence but is unable to produce a functional product.

Reverse transcriptase: An enzyme that directs DNA synthesis from an RNA template. The process is called reverse transcription and is part of the replication cycle of certain RNA viruses.

Representations of Protein Patterns From Two-Dimensional Gel Electrophoresis Databases

Peter F. Lemkin

Two-dimensional polyacrylamide gel electrophoresis is a technique for separating hundreds to thousands of polypeptides by two distinct characteristics: Isoelectric point and apparent molecular mass. One goal in analyzing a group of gels is to find interesting sets of polypeptides which appear as spots. A second goal is to make sense of this data when large numbers of spots, gels, and experimental conditions are involved. As part of the GELLAB-II system, we can present this data in different ways or "views". We suggest that these tools may be used to help find co-regulated genes.

Introduction

Two-dimensional polyacrylamide gel electrophoresis is a technique which can separate mixtures of hundreds to thousands of polypeptides by two distinct characteristics: isoelectric point and apparent molecular mass.[1] The problem then becomes one of comparing 2-D gels from different samples run under different conditions. For comparing two gels with only a few polypeptide differences, one could slide one gel over the other on a light box (for back illumination) until corresponding spots are aligned in particular regions of the gels. Because of "rubber sheet" distortions created when 2-D gels are run, spots will generally not line up perfectly in all parts of the gels for a given global alignment. Even more unwieldly is the visual comparison of many changes between several gels.

We have developed a multiple 2-D gel image analysis and exploratory database system called GELLAB-II[2-5] for analyzing these gels. It finds all spots in all gels, pairs spots between each gel and a reference gel, and merges these paired spot data into a single "3-D" gel composite database. The composite database is illustrated in Fig. 1. The database is constructed around a

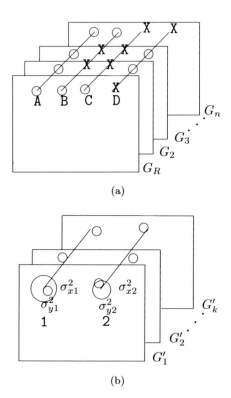

Fig. 1. 3-D composite gel database model. (a) illustrates a composite gel database. Corresponding paired spots (circles) are denoted by diagonal lines drawn through them. Such sets of corresponding spots are called *Rspot* sets. One of the gels is selected to be a reference or *Rgel*, denoted G_R. The circle means the spot is present and the X means that it is missing in that gel. (b) shows the basis of using mean spot positions for estimating *canonical* spots for a subset of k gels from the n gel database. The canonical spot may then be used to estimate the position of spots missing from some of the other gels.

Reference gel (any good gel from the set of gels used in the database). Groups of corresponding spots are called *Rspot* sets of spots. Note that these sets in this database are assigned sequential Rspot numbers which indicate a set of corresponding spots from different gels. The database, then, consists of all Rspot sets which contain a spot found in one or more gels.

In Fig. 1(a), spot A occurs in all n gels. Spot B occurs in the Rgel and in one other gel. Spot C is only in the Rgel. Spot D is not present in the Rgel but is in most of the other gels. Spots A, B, and C are in the un-extended Rspot database (since they occur in the Rgel) while spot D is in the extended or *eRspot* part of the database (since it does *not* occur in the Rgel). In Fig. 1(b), the mean and variance (σ_x^2, σ_y^2) of Rspot positions across a set of gels is mapped to the coordinate system of the Rgel. When that has been done, the set of gels of the same experimental class may be replaced by a single averaged gel called the *Cgel'* (the estimate of the canonical gel). The mean displacement vector of a canonical spot from its associated landmark spot (in any gel under discussion) is used to extrapolate the position in gels where the expected canonical spot is missing.

This database may be explored in a huge number of ways to discover patterns of protein changes. These include various parametric and non-parametric tests to find spot differences which correspond to changes in the experimental variables. The relationship between protein concentration and spot integrated density is approximately stochiometric (approximately linearly related), and so we define protein concentration in terms of integrated spot density corrected for background. Because of systematic variability between gels, gel density data from different gels is normalized before using it in calculations.

We look for protein spots changing as a function of experimental conditions. We also look for subpatterns of spots which are changing together in what may be candidates for co-regulation. That is, polypeptides (gene products) which are induced or decreased simultaneously, may have their genes, which direct their production, modulated by another common gene or common gene expression pathway. This paper shows some of the derived image and graphic representations for viewing these patterns in various physical and abstract slices of this "3-D" gel database — which we might consider to be a 3-D spreadsheet for gels. As part of the GELLAB-II system, we may present this gel data in different ways or "views". Our graphic methods are divided into derived images, schematics, graphics, and tables.

Derived Images

Often when working with data derived from images one wants to see marks
on the object or objects under discussion. The following three displays help
make this data object to image connection. The **Rmap image** in Fig. 2 is an
overlay of a particular gel image with numeric names from the 3-D database
of a particular *set* of Rspots. These spots were found in a preliminary exploratory analysis by an initial t-test search of the 29-gel database for quanti-

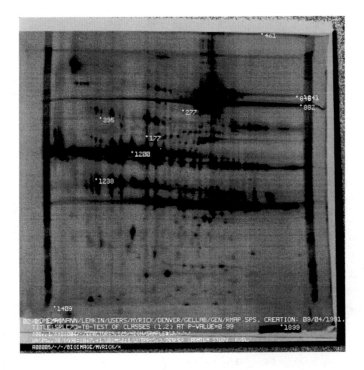

Fig. 2. The **Rmap image** is an overlay of a *particular* gel image with numeric names of
a particular set of Rspots. The region of interest may be magnified if several Rspots are
labeled in the same region. Colored labels may be used to indicate different conditions. An
Rmap is useful for indicating where Rspots are located which are statistically interesting
and also for finding *false positive* test results by manual inspection. This Rmap is of a silver
stained gel of human urine from an ongoing cadmium toxicity study by James Myrick of
the Centers for Disease Control in Atlanta. We are grateful to Dr. Myrick for providing us
with this gel data. (The actual grayscale image has better resolution than this laser printer
version.)

tative differences between persons with "low" and "medium" urinary cadmium concentration gels ("high" concentration data is not shown). Inspection revealed possible artifacts from edge effects (Rspots 461, 841, 845, 882, 1409 and 1899) (*false positive* spots) as contrasted with others of possible interest (Rspots 177, 277, 395, 1200 and 1238).

The **mosaic image** in Fig. 3 consists of subregions from different gels in a database surrounding a particular Rspot in those gels. In gels where the

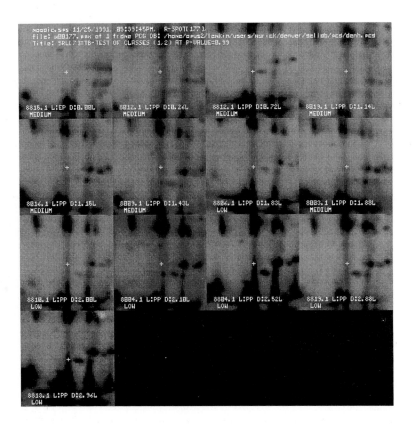

Fig. 3. The **mosaic image** consists of subregions of different gels in a database surrounding a particular Rspot over that set of gels. The panels are sorted top to bottom, left to right by protein concentration. They are labeled with gel name, experimental class, and protein density (concentration). The mosaic is of Rspot[177] for a subset of the gels in the cadmium toxicity study shown in Rmap of Fig. 2. (The actual grayscale image has better resolution than this laser printer version.)

spot is missing we extrapolate its expected position. The mosaics are used for several purposes to:

(1) verify two-condition statistical tests where the spot is larger and darker in one class of gels than in the other;

(2) verify *False*[+] test results for mispaired spots; and

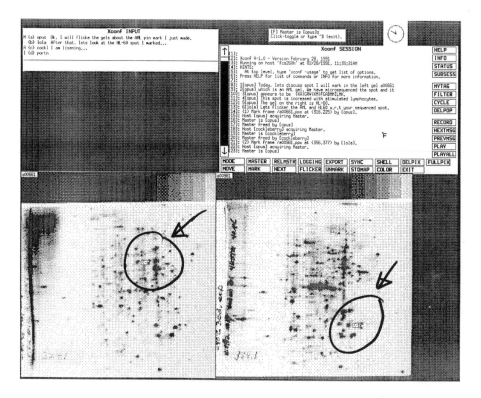

Fig. 4. The **pin-marked image** is similar to the Rmap image where colored labeled pins are inserted by several *different* investigators. Users interact with the same image(s) using wide-area network-based image-conferencing software such as our Xconf system.[6] This screen dump shows a four-user conference with color-coded pin marks in the human lymphocyte 2-D gel images indicating spot objects of interest. (The pin-marks are located inside the hand drawn circles.) Other operations such as image-flickering, movies, etc. may also be performed and they are visible simultaneously to all members of the conference. Asynchronous conversations between users can take place in the Xconf-INPUT window while conference dialog and actions are recorded in the scrollable Xconf-SESSION window. Conference control may be passed among users so different members may try out their ideas on the conference.

(3) verify *False*⁻ test results for spots which "make sense" but were mis-paired. The latter is done, for each Rspot, by extrapolating the positions of spots missing from some of the gels.

The **pin-marked image** shown in Fig. 4 is similar to the Rmap image in indicating objects in the image. Here, colored labeled pins may be inserted simultaneously by several *different* investigators interacting with the same image through network-based *image-conferencing* software.[6]

Schematic Plots: Rmaps and Mosaics

An alternative representation of the derived images are the Rmap and mosaic plots shown in Figs. 5(a) and 5(b). The advantage of the plots is that they are sometimes easier to read and photocopy — the disadvantage is they lose the contextual background information surrounding spots in the image views. The advantage of the schematic plots is that they are sometimes easier to read and photocopy, the disadvantage is that they lose the contextual information surrounding spots such as gel background, streaks, etc. Empty panels appear in the mosaic plot for gels not in this subset of the database.

Graphic Representations

For a given protein Rspot, we may observe its mean abundance $\overline{D}$ with respect to a set of different gels from different experimental conditions. $\overline{D}$ is computed by averaging integrated density corrected for background for all corresponding spots in an Rspot. A set of n observations $\{\overline{D}_1, \overline{D}_2, \ldots, \overline{D}_n\}$ is called an *expression profile* for n experimental conditions. The next stage of analysis is to see which Rspots have similar expression profiles. Proteins which exhibit co-regulation effects would be expected to have similar expression profiles.

The **expression-profile plots** in Fig. 6 displays all the protein expression profiles of a set of proteins as a function of experimental conditions. However, this does not explicitly show which Rspots are most similar. Standard clustering techniques may be used with 2-D gel data. The **Rspot clustering dendrogram** groups similar Rspots "objects" from a set of Rspots as a function of protein expression profiles "features". Proteins which may exhibit co-regulation effects would be expected to have similar expression profiles. By

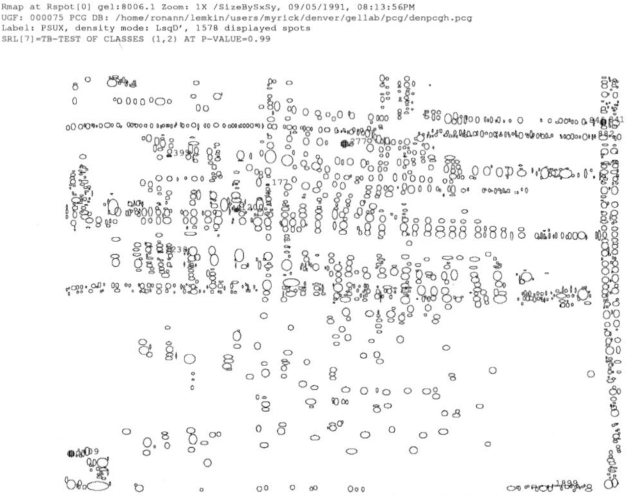

Rmap at Rspot[0] gel:8006.1 Zoom: 1X /SizeBySxSy, 09/05/1991, 08:13:56PM
UGF: 000075 PCG DB: /home/ronann/lemkin/users/myrick/denver/gellab/pcg/denpcgh.pcg
Label: PSUX, density mode: LsqD', 1578 displayed spots
SRL[7]=TB-TEST OF CLASSES (1,2) AT P-VALUE=0.99

(a)

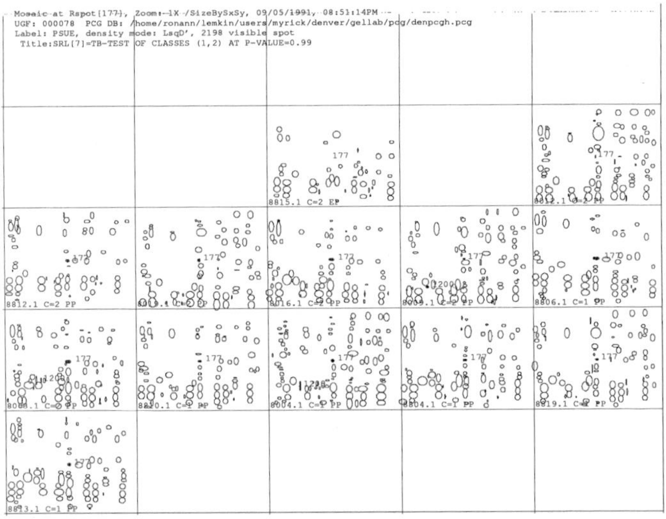

Mosaic at Rspot[173], Zoom:1X /SizeBySxSy, 09/05/1991, 08:51:14PM
UGF: 000078 PCG DB: /home/ronann/lemkin/users/myrick/denver/gellab/pcg/denpcgh.pcg
Label: PSUE, density mode: LsqD', 2198 visible spot
Title:SRL[7]=TB-TEST OF CLASSES (1,2) AT P-VALUE=0.99

(b)

Fig. 5. An alternative representation to derived images are the (a) **Rmap** and (b) **mosaic** schematic plots (of the cadmium toxicity data from Fig. 2). The shape of a spot, for a particular gel, may be plotted as either that of the original gel spots or ellipses whose size is proportional to protein concentration. In the former case, increasing density is indicated by increased width of cross-hatched bars inside of labeled spots.

Expression Profile - Density vs Class Values 09/10/1991, 03:02:58PM

TITLE: SRL[6]=TB-TEST OF CLASSES (1,2) AT P-VALUE=0.95
CLASS 1: LOW
CLASS 2: MEDIUM
CLASS 3: HIGH

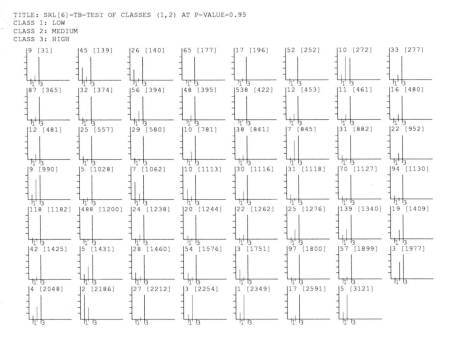

Fig. 6.　The **expression-profile plots** displays the protein expression profiles of a set of proteins on the Y axis as a function of experimental conditions on the X axis (data from cadmium database discussed in Fig. 2). A normalized expression profile of a Rspot for n experimental conditions with mean expressions $m_1, m_2, \ldots, m_n$ is an n-tuple $(1.0, m_2/m_1, \ldots, m_n/m_1)$.

placing a large number of expression profiles in one visual space, one may see trends. However, it is up to the observer to make those connections.

Figure 7(a) shows expression profiles of 13 proteins previously determined to be modulated by environmental influences while Fig. 7(b) shows a cluster analysis of spots as a function of their expression profiles.[7] In Fig. 7(a), the expression profiles were manually clustered according to their most prominent feature. Heterogeneity of these groups is obvious at more detailed inspection and has been dealt with by computerized cluster analysis (see Fig. 7(b)).

(i) Cluster I proteins: Proteins with high relative abundance in the absence of any co-cultured cells (none).

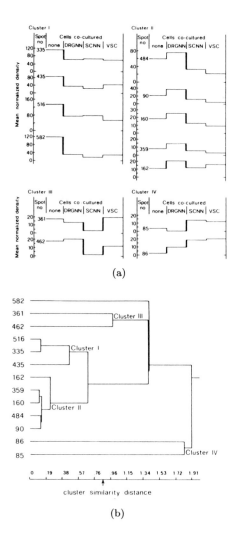

(a)

(b)

Fig. 7. Dendrogram used to **cluster spots** as a function of their expression profiles. (a) Expression profiles of 13 proteins previously determined to be modulated by environmental influences. Histogram-type plots of mean normalized density versus co-culture condition. The thick vertical bars indicate where statistically significant differences in the comparison of the adjacent experimental classes occur. (b) Dendrogram representing the clustering pattern of the 13 environmentally modulated proteins. Computerized cluster analysis was performed with the 13 modulated spots as the objects and the ratios of the mean values of all permutations of pairwise comparisons of all experimental classes as the features. (With permission from *Developmental Biology*.)

(ii) Cluster II proteins: Proteins with high relative abundance under local co-culture with peripheral non-neuronal cells (DRGNN).

(iii) Cluster III proteins: Proteins with depressed relative abundance under co-culture with central non-neuronal cells (SCNN).

(iv) Cluster IV proteins: Proteins with high relative abundance under local co-culture with central nervous system cells (SCNN and VSC).

In Fig. 7(b), the numbers to the left of the tree indicate the two-dimensional-gel Rspot numbers. The abscissa represents cluster similarity distance as measured in standard error distance. As a cutoff criterion of "significant

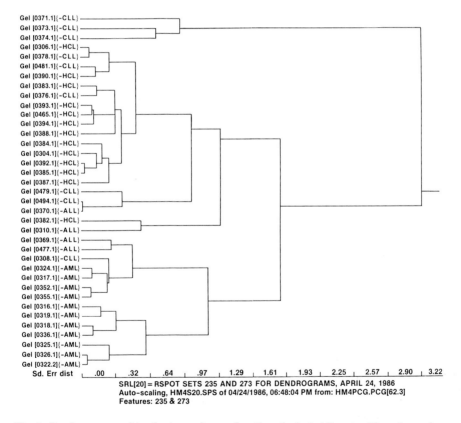

SRL[20] = RSPOT SETS 235 AND 273 FOR DENDROGRAMS, APRIL 24, 1986
Auto-scaling, HM4S20.SPS of 04/24/1986, 06:48:04 PM from: HM4PCG.PCG[62.3]
Features: 235 & 273

Fig. 8. Dendrogram used to **cluster gels** as a function of selected Rspots. The gels are from our HM5 leukemia database as a function of myeloid marker proteins 235 and 273. (With permission from *Electrophoresis*.)

clusters" the mean cluster similarity distance plus one standard deviate was used (cutoff indicated by vertical arrow).[7]

Similarly, we may cluster separate gel samples as a function of similarity of *sets* of key proteins as shown in Fig. 8. Note the acute myelogenous leukemia (AML) gels clustered together as would be expected using marker proteins for that class. However, the preliminary data for acute lymphoid leukemia (ALL) shows that it is split between two major clusters suggesting possible bimodality within the ALL data. The other lymphoid classes are chronic lymphoid leukemia (CLL) and hairy cell leukemia (HLL). This technique may be used for partioning gels based on protein expression similarity.[4] Although not visible here, gels names in each experimental class (AML, ALL, CLL and HCL) are different colors to aid in observing the clustering's validity.

We have also found the *N*-**gel scatter plot** suggested by Cleveland[8] to be useful (not shown). Here, N is the number of experimental conditions and each scatter plot is a density (protein concentration) versus density plot. The scatter plot points are each spot found in the corresponding gels used in the particular scatter plot. These subplots are assembled as a upper-diagonal matrix for each of the experimental conditions so that similar scatter plots patterns may be visually identified.

"Tabular" Plots

Although not exactly graphs, there are several other quasi-graphical displays which represent data dependence by Rspots' positions relative to one another. Think of these forms as histograms where the spot's name goes into the histogram bins.

The **change histogram** in Fig. 9 displays Rspot numbers in a histogram as a function of the ratio of mean protein expressions for two experimental conditions. Rspots clustered near particular bins may be considered as possible candidates for co-regulation mechanisms. To make it easier to view, we collapse empty bins of the histogram when there is nothing there for 3 bins on either side of an entry. Computing change histograms for experimental conditions other than those used in originally finding a set of Rspots may be useful for finding additional subpatterns with other gels in that same set of Rspots.

The **ordered expression-profile table** shown in Fig. 10 is a table showing which Rspots are most similar in a given set of Rspots. It lists those below a threshold cutoff indicating the most similar expression profiles. Closest Rspots

```
                          Change histogram
-----------------------------------------------------------------
m2/m1 Rspot sets
 0.20 882
 0.25
 0.30
 0.35
 0.40
 0.45 177
 0.50
 0.55
             .
             .
 3.75
 3.80 277
 3.85
 3.90
 3.95
 4.00 841
 4.05 990
 4.10
 4.15
             .
             .
 4.80
 4.85 395
 4.90
 4.95
             .
             .
 5.25
 5.30 1200
 5.35
 5.40
 5.45
 5.50
 5.55 461
 5.60
 5.65
 5.70
 5.75
 5.80 1238
 5.85
 5.90
             .
             .
 7.00
 7.05 1899 2212
 7.10
 7.15
             .
             .
 8.95
 9.00 1409
 9.05
 9.10
             .
             .
 9.95
10.00 845
```

Fig. 9. The **change histogram table** displays Rspot numbers for any two experimental classes in a histogram as a function of the ratio of mean protein expression for two experimental conditions (data from Fig. 2). The bins cover the range of [1/20:20/1].

```
Order Expression Profile of Rspots in SRL.

Mean-Dens-Class-I/Mean-Dens-Class-J for Min-LsqErr-threshold=5.00
Rspot:   m1/1 m2/1 m3/1
-----------------------------------------------------------------
   177     1.0   0.5 27.2
   277     1.0   3.8 19.2
   395     1.0   4.9 25.1
   461     1.0   5.6 24.4
   841     1.0   4.0 22.3
   845     1.0  12.6 16.0
   882     1.0   0.2  5.8
   990     1.0   4.1  4.9
  1200     1.0   5.3+99.9
  1238     1.0   5.8 30.3
  1409     1.0   9.0 25.0
  1899     1.0   7.1 82.9
  2212     1.0  19.8+99.9

Similar EPs Sorted by Minimum lsqErr of profiles (Rspot#:lsqErrValue)
-----------------------------------------------------------------
   177     395:3.45  461:4.12  841:4.31  1238:4.37
   277     841:2.20  461:3.90  395:4.27
   395     461:0.70  841:2.09  1409:2.96  177:3.45  1238:3.75  277:4.27
   461     395:0.70  841:1.86  1409:2.50  277:3.90  177:4.12  1238:4.19
   841     461:1.86  395:2.09  277:2.20  1409:4.05  177:4.31
   845
   882     990:2.82
   990     882:2.82
  1200
  1238     395:3.75  461:4.19  177:4.37  1409:4.39
  1409     461:2.50  395:2.96  841:4.05  1238:4.39
  1899
  2212
```

Fig. 10. The **ordered expression-profile table** shows Rspots in a set of Rspots which are most similar to each other (data from Fig. 2). It computes the expression profiles for a selected set of Rspots, and then computes the least square error for each Rspot's expression profile with all of the others.

(i.e., most similar) are listed first on the branches of this "tree". The least square similarity measure (*lsqErr*) is also included after each Rspot in the list to give additional information. (Entries with values > 99.9 are indicated by $+99.9$.)

Another way of plotting individual gel expression profiles is the **Rspot rank-order table** in Fig. 11 which plots the density values of spots from individual gels for a few Rspots. The individual spot entry, e.g., 8006.1P3, is decomposed as follows. A gel is referenced in GELLAB-II by its accession number (e.g., 8006.1). Spots are paired with the reference gel with different pairing quality labels: S is sure-pair, P is possible-pair, and all data in this table are possible-pairs. Finally, the experimental class that the gel belongs to is specified by the experimental class number which is listed at the bottom of the table. By adding connecting lines for the same gels with different Rspots

```
RANK-ORDER table: <ACC#>&<pairing-label>&<Class #>

Density
  118.1 |  8006.1P3
  115.1 |  8808.1P3
  112.2 |
  109.2 |
  106.3 |
  103.3 |  8001.1P3
  100.4 |                              8809.1P3
   97.4 |                 8001.1P3
   94.5 |
   91.5 |
   88.6 |
   85.6 |
   82.7 |
   79.7 |
   76.8 |
   73.8 |
   70.9 |
   67.9 |
   65.0 |                 8006.1P3
   62.0 |
   59.1 |
   56.1 |
   53.2 |
   50.2 |
   47.3 |  8809.1P3
   44.3 |
   41.4 |  8801.1P3
   38.4 |
   35.5 |
   32.5 |
   29.6 |
   26.6 |                 8801.1P3
   23.7 |                              8006.1P3
   20.7 |
   17.8 |                              8001.1P3
   14.8 |                 8809.1P3
   11.9 |  8816.1P3
    8.9 |                 8808.1P3 8815.1P2
        |                 8011.1P2
    6.0 |  8807.1P3 8009.1P2 8802.1P2
        |                 8012.1P2
    3.0 |                 8820.1P3 8004.1P1
        |                 8019.1P2
        |                 8805.1P1
        |                 8812.1P2
        |                 8806.1P1
    0.1 |  8813.1P1 8816.1P3 8813.1P1
        |  8819.1P1 8819.1P1 8005.1P1
        |  8804.1P1 8004.1P1 8018.1P1
        |  8004.1P1 8804.1P1
        |  8810.1P1 8813.1P1
        |  8003.1P2
        |  8806.1P1
        |  8009.1P2
        |  8016.1P2
        |  8019.1P2
        |  8812.1P2
        |  8012.1P2
        ---------------------------------------
Rspot#      177        277       395
Class #  1=LOW
Class #  2=MEDIUM
Class #  3=HIGH
```

Fig. 11. The **Rspot density rank-order plot** plots the density values of spots from individual gels (data from Fig. 2). The individual spots from particular gels are indicated by numbers (e.g., 8006.1P3, 8809.1P3, etc.) and are drawn in rank-order in the plot with normalized spot density D on the vertical axis. Their corresponding Rspot numbers (177, 277 and 395) are on the horizontal axis.

(use of different colors is strongly advised), it is useful for comparing protein expression of several different Rspots. Outlier spots or gels are easily identified since the lines or curves are not parallel to those of other gels. For example, the individual spot entry, e.g., 8006.1P3, is decomposed as follows. A gel is referenced in GELLAB-II by its accession number (e.g., 8006.1). Spots are paired with the reference gel with difference pairing quality labels: S is sure-pair, P is possible-pair, and all data in this table are possible-pairs. Finally, the experimental class that the gel belongs to is specified by the experimental class number which is listed at the bottom of the table. By adding connecting lines for the same gels with different Rspots (use of different colors is strongly advised), it is useful for comparing protein expression of several different Rspots. Outlier spots or gels are easily identified since the lines or curves are not parallel to those of other gels.

Summary

When analyzing large numbers of spots from large numbers of gels it is possible to shrink the problem by successive data reductions and to draw conclusions along the way by visually comparing partial results. Significant visual patterns in some views of the reduced data are often more apparent than others and may suggest further experiments to be performed on eluted proteins found by such an exploratory data analysis. The unifying theme of these graphical methods is to make it easier to view the same data in different ways — perhaps gaining new insights with these different views. The GELLAB-II system is available for research groups via a Material Transfer Agreement with NCI and our Laboratory. More information on GELLAB and Xcom can be obtained from our World Wide Web server with URL at http://www-lmmb.ncifcrf.gov/.

Acknowledgements

Thanks for constructive suggestions for this paper from Tom Schneider, Kyle Upton, and Jim Myrick. Thanks also to Jim for use of some of his data used in many of the examples.

References

1. O'Farrell, P.H., "High resolution two-dimensional electrophoresis of proteins", *J. Biol. Chem.* **250**, 4007–4021 (1975).

2. Lipkin, L.E. and Lemkin, P.F., "Database techniques for multiple PAGE (2-D gel) analysis", *Clin. Chem.* **26**, 1403–1413 (1980).

3. Lemkin, P.F. and Lipkin, L.E., "Database techniques for 2-D electrophoretic gel analysis", *Computing in Biological Science*, eds. M. Geisow and A. Barrett (Elsevier/North-Holland, 1983) pp. 181–226.

4. Lemkin, P.F. and Lester, E.P., "Database and search techniques for 2-D gel protein data: A comparison of paradigms for exploratory data analysis and prospects for biological modeling", *Electrophoresis* **10**(2), 122–140 (1989).

5. Lemkin, P.F., "GELLAB-II, A workstation based 2-D electrophoresis gel analysis system", *Two-Dimensional Electrophoresis*, eds. T. Endler and S. Hanash (VCH Press, 1989) pp. 53–57.

6. Lemkin, P.F., *XConference — A Multimedia Image Conferencing System* (1991).

7. Sonderegger, P., Lemkin, P.F., Lipkin, L.E. and Nelson, P.G., "Coordinate regulation of the expression of axonal proteins by the micro-environment", *Developmental Biology* **118**, 222–232 (1986).

8. Cleveland, W.S., *The Elements of Graphing Data* (Wadsworth Advanced Books and Software, 1985).

Glossary

***coregulated genes*:** Polypeptides (gene products), which are induced or decreased simultaneously, may have their genes which direct their production modulated by another common gene or common gene expression pathway.

***image conferencing*:** Networked-based image display and manipulation software which enables a number of participants at different sites to interact with an image simultaneously viewed by all participants.

***isoelectric point*:** The point (pIe), when running a polypeptides in a one-dimensional gel in a pH gradient gel, where each polypeptide reaches the state where its inherent pH balances the pH of the gel resulting in no net charge is called its isoelectric point. In a 2-D gel, this is generally presented as the horizontal dimension with low pIe on the left.

***polypeptides*:** A linear chain of amino acids. Proteins may contain one or more polypeptides.

***rubber sheet distortion*:** The composite two-dimensional surface created by a number of local magnification (minification) over the surface of an object. Similar to when one draws a face on a balloon and then stretches in several directions at the same time. A similar effect occurs with 2-D gels which must be taken into account when different gels are compared.

stochiometric: The approximately linear relationship between a quantity being measured and the quantity of the material that it represents. For example, protein concentration and spot integrated density from gel autoradiograph films is approximately stochiometric under ideal conditions.

A Protein Visualization Program

D.A. Kuznetsov
and
H.A. Lim

The power of today's graphics workstations has made visual presentation of scientific results commonplace. Often the results presented do not require complicated analysis, and consequently, the use of complex software packages is not necessary. Thus, simple and easy-to-use programs are desirable. This paper reports on a simple program, VisiCoor, which can be used for the visualization of protein structures.

Introduction

Though many commercial protein visualization packages exist on the market,[1] users can at times encounter difficulty when attempting to modify the algorithm so that particular features of the biological molecule under study can be highlighted. VisiCoor (**Visi**ble **Coor**dinates)[2] is an easy-to-use and handy program which currently exists in two versions: One for the VAX-Tektronix 4129 combination and one for the IRIS 4D/240GTX. It is non-proprietary, thus offering users more flexibility than typical commercial packages. It has a flexible logic system consisting of a large set of basic rules and supports high quality raster graphics. Through a user-friendly interface, users can manipulate the program in such a way that unnecessary details of the protein structure are suppressed, thereby enabling them to concentrate on substructures of immediate interest. This relieves users from information overload and allows them to capitalize on the power of human recognition of images.

Features of Visicoor

The complete electronic data specifying protein coordinates are stored in database centers such as the Brookhaven Protein Data Bank.* Often the data contains more information than is needed for analysis. To view a subset of atoms of interest, a Boolean flag is defined through a mask

$$M(i; \alpha, \ldots, \delta) = \left\{ \begin{array}{ll} .\text{TRUE.} & \text{if criteria satisfied for atom } i \\ .\text{FALSE.} & \text{otherwise}, \end{array} \right.$$

which, when applied to the database center supplied protein data, selects those atoms satisfying the criteria $\alpha, \ldots, \delta$. The mask functionally marks only pertinent atoms as "active" (i.e., $M(i; \alpha, \ldots, \delta) = .\text{TRUE.}$). It enables users to manipulate these logically linked atoms as a single object or a class.

Class building

The above process of selecting pertinent atoms for visualization is called *class building*. The program has twelve preset criteria $(\alpha, \ldots, \delta)$. However, if users elect to create their own list of criteria, they can do so by having the program read in a user-supplied criterion list from an externally provided file. The preset criteria are categorized based on structural, physico-chemical, and geometrical properties of fragments of the protein under investigation. In the "Structure" category, users can either choose residues or atoms which belong to the main chain, α helices, or β sheets, or they can choose turns or atoms which have the selected range of values of torsion angles ϕ–ψ. An illustration of the former is shown in Fig. 1 and two examples of the latter are given in Figs. 2 and 3. In the "Residue" category, users can choose residues which are hydrophobic, aromatic, charged, by name or by specifying a field (i.e., from residue i to residue j). In the "Geometry" category, users can choose atoms which are in the proximity of a particular atom as depicted in Fig. 4 or they can choose atoms which have no bonds to any other atoms.

---

*For contact: Protein Data Bank, Chemistry Department, Building 555, Brookhaven National Laboratory, Upton, New York 11973, USA, Tel: (516) 282-3629; Fax: (516) 282-5815; E-mail: pdb@bnlchm.bitnet.

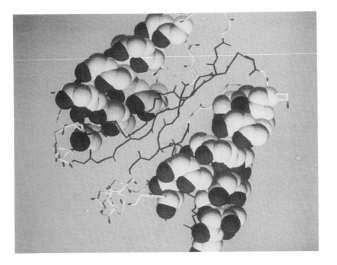

Fig. 1. An illustration to show *class building* based on the structural properties of the main chain of protein 4FXN. The four α helices are covered by van der Waal spheres and the other substructures (β sheets and irregular structures) are represented by matchsticks.

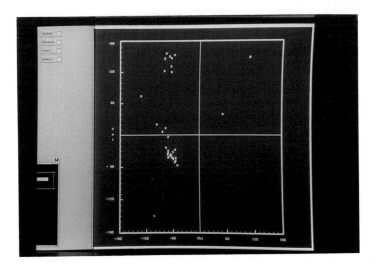

Fig. 2. An example to show *class building* based on the structural properties (ϕ–ψ) of the protein 1PPT. In this example, only the map is depicted. Note that the crowded points in the neighborhood of (ϕ, ψ) values corresponding to (-60, -60) have been selected for plotting Fig. 3.

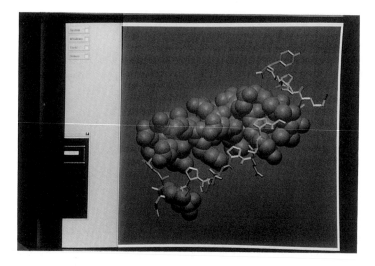

Fig. 3. The image of the protein 1PPT of Fig. 2. To create this image, the cluster of points nearby (ϕ, ψ) values of $(-60, -60)$ are first selected and then dressed by balls.

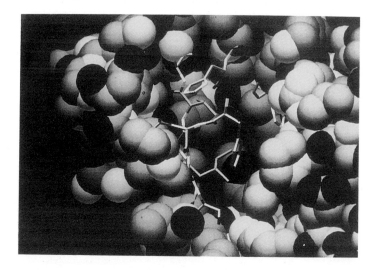

Fig. 4. An example to illustrate the use of the geometrical properties of the active site of the protein 1RN3 in *class building*. The atoms which are within the intersection of the protein globule with the two spheres centered on the ND1 of His119 and the CG of Phe120 are represented by matchsticks. Other atoms are represented by balls. The CG of Phe120 is at the upper right in the center of the image.

There are four ways to accomplish the selection of pertinent atoms:

$$A \rightarrow \begin{cases} \overline{A}, & \text{(inversion)}; \\ B, & \text{(overwriting)}; \\ A \cup B, & \text{(addition)}; \\ A \cap \overline{B}, & \text{(deletion)}, \end{cases}$$

where A is the current class, B is the newly created class. Before A is modified, it is always copied into a buffer so that the user can recover it if so desired.

Graphics

Having built the *class*, users can then proceed to display the image on graphics terminals.

VisiCoor uses the conventional atomic representations: A *stick* to emulate a covalent bond, a *ball* to emulate the van-der-Waals space of an atom, a combination of *ball-and-stick* (functionally the same as *stick*) and *dot-shell* (semitransparent *ball*). The first three representations are constructed from colored, 3-source lighted, Gouraud-shaded, z-buffered, depth-cued facets in the Tektronix version of VisiCoor and from colored, 3-source lighted, Gouraud-shaded, z-buffered, "polished" facets in the Iris version of VisiCoor. The fourth representation consists of color points on a spherical shell. By default, all atomic representations are colored using the CPK-palette convention: H — white, C — grey, N — blue, O — red, S — green, P — yellow. However, users can change this default color scheme. The colors violet, orange, and yellow are available for this purpose.

Though the conventional shape for the atomic representation of a *stick* is a long cylinder, a long parallelepipe ("matchstick") is used in VisiCoor. The "matchstick" consists of two joined halves to represent the two contiguous atoms. Each half is colored according to the chemical types of the contiguous atoms. The calculation of the vertex coordinates of a "matchstick" is not obvious because the angle of rotation of a "matchstick" about an atom-atom axis is ambiguous and cannot be determined uniquely solely from the pair of atomic coordinate vectors. This problem is solved by an original algorithm implemented in the program. The algorithm constructs a "matchstick" in a manner so that the two short edges of each side of the "matchstick" lie in the x-y plane. Some of the advantages of using a "matchstick" instead of a cylinder are: (1) Oriented sticks reduce the possibility of confusion and help the viewer to perceive the image as a structural object. This aids in the estimation of the direction of a bond or of the dihedral angles between atoms, as illustrated in Fig. 5. (2) In the construction of a *stick*, two factors help improve the computational performance. First, the use of flat edges turns off the Gouraud interpolation of colors within an edge. Gouraud interpolation cannot be omitted in the cylinder representation. Second, the use of a reduced number of long edges decreases the computational time. The improvement in computational performance makes possible an interactive access by users to the colored-, lighted-, z-buffered-image of a protein up to as large as thousands of atoms.

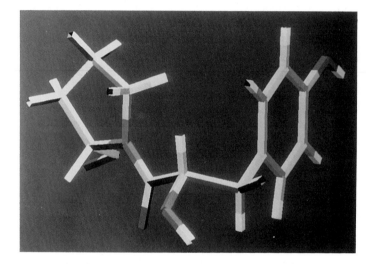

Fig. 5. Dipeptide Tyr-Pro is used as an example to show the advantage of using a "match-stick" over a "cylinder" in atomic representations. The edges on the matchsticks help the viewer to estimate the direction of bonds as well as torsion angles between atoms.

The *ball* primitive is approximated by two-polar polygons (like domes). To enhance performance, only the foreground semispheres are displayed. The *ball-and-stick* primitive consists of the *stick* in combination with a scaled down *ball*. The *dot-shell* consists of points lying on atomic van-der-Waals spheres. Points which are in the background and thus obscured by other *dot-shell* spheres are automatically removed by the program. The atoms selected in the *class building* can be represented by any of the four representations or combinations thereof.

The *stick-*, *ball-*, and *ball-and-stick*-represented images are interactively rotatable. The speed and axis of rotation can be controlled. Real-time manipulations of images represented by *sticks* are possible with molecules consisting of up to a thousand atoms. However, operations involving images represented by *balls* can only be performed interactively on a more modest molecular size of up to fifty atoms. The *dot-shell* (if applied) disappears during rotation and is recalculated automatically for the new viewing angle.

Input data file format

The program assumes that the input files are in the Protein Data Bank format.[3] Certain fields of the PDB format (HELIX, SHEET, TURN, HETATM) are essential for the logic features of the program. In the preface menu the following options are available:

1. Read all atoms, the main chain, or C_α atoms.
2. Read or skip heteroatoms.
3. Set up the fragments to be read on the basis of residue numbers (up to 20 fragments).
4. Set new covalent radii of C, H, N, O, P, S atoms.

User interface

The user interface is written as a set of hierarchical pop-up menu frames. The nine menu items supported are: Help, Rotate, Dress, Set-class, Paint, Calc, List-class, Erase, Setup.[2]

In both versions of the program the mouse is used as a means of interactively polling the image. By clicking on a particular atom, the user may get a report about the atom and its residue number. All time-consuming operations are accompanied both by an indicator bar with the length of the bar representing the fraction of the operation completed, and by a descriptive message.

As an example, the following menu options may be used in creating Figs. 2 and 3 on an IRIS 4D/240GTX graphics workstation:

1. Start up VisiCoor on the IRIS.
2. At the prompt for input data file, type in the input data file name. In this case, the input data is pdb1ppt.ent in PDB format.

3. When the menu options pop up, select "SELECT ATOMS...***" where "***" can be either "ALL", "MAIN CHAINS", or "Ca ONLY". These options can be obtained by hitting the carriage return key while selecting "SELECT ATOMS...***". Once the proper "SELECT ATOMS...***" has been selected (and other options default), hit the up arrow key back to "BEGIN READING". The program will then start reading in the atomic coordinates. While this is being performed, the total number of atoms read will be flashed on the screen.

4. "DRESS" the protein with *sticks*.

5. "PAINT" the protein yellow. This color is chosen for a good contrast in a black and white photo plate. Other color selections can be chosen as well.

6. Under "SET_CLASS" and "STRUCTURE", select "BY PHI-PSI". A map of points will appear on the screen. There is a crowded region nearby $\phi = -60$ and $\psi = -60$. This region is marked by clicking the mouse on the upper left corner of the enclosing rectangle and releasing the mouse button at the lower right corner of the enclosing rectangle.

7. Complete the "CLASS BUILDING" step by selecting "EXIT, class setting okay".

8. "DRESS" the class with *balls*.

9. "PAINT" the *balls* orange. Other color selections may be used.

10. If so desired, the image can be rotated to an optimal viewing angle by depressing a mouse button. The speed of rotation is proportional to the distance from the center of the image and the axis of rotation is perpendicular to the vector joining the center of the image to the point of the cursor arrow.

Conclusion

Two versions of the molecular graphics program VisiCoor, one for the VAX-Tektronix 4129 and one for the IRIS 4D/240GTX Silicon Graphics workstation,[4,5] are described. The VAX-Tektronix 4129 combination is a VAX computer that displays its output on a Tektronix graphics terminal and the IRIS 4D/240GTX Silicon Graphics workstation is a computer capable of displaying its own output using integrated, high-speed graphics hardware. These workstations are widely used for scientific visualizations. The program, which has full graphics capabilities, is used for displaying protein structures. Sample plots drawn using the program's ability to manipulate the display of protein structures can produce very instructive graphics for visualizing the supersecondary structures, functional sites, and other features of proteins. The program may be particularly useful in those investigations of protein structures where the description cannot be easily provided by mathematical approaches, but where human intuition and knowledge might provide a better guide in their interpretation.

Interested readers may write to the authors to request for a free copy of the VisiCoor program which will run on the console of an IRIS 4D/240GTX. For a VAX-Tektronix 4129 version, readers may have to write to Russia.

Acknowledgements

The original program was written for the VAX-Tektronix 4129 at the Engelhardt Institute of Molecular Biology during 1988–1989. The program was ported to the IRIS 4D/240GTX workstation at the Supercomputer Computations Research Institute (SCRI) of the Florida State University in 1990. HAL is partially funded by TRDA grants under Contract Numbers TRDA 116 and TRDA 205, and by SCRI which is partially funded by the US Department of Energy under Contract Number DE-FC05-85ER250000. HAL would like to thank the hospitality of the Engelhardt Institute of Molecular Biology (Academicians A.A. Bayev and A.D. Mirzabekov) and the Keldysh Institute of Applied Mathematics (Academician S.K. Kurdyumov) during his visit to the Russian Academy of Sciences, and DAK would like to thank SCRI for the hospitality extended to him during his visit to the US. These exchanges have made the project possible.

References

1. Brooks, B.R., Bruccoleri, R.E., Olafson, B.D., States, D.J., Swaminathan, S. and Karplus, M., "CHARMM: A program for macromolecular energy, minimization, and dynamics calculations", *J. Computational Chemistry* **4**(2), 187–217 (1983).
2. Kuznetsov, D.A. and Lim, H.A., "VisiCoor — A simple program for visualization of proteins", *J. Molecular Graphics* **10**, 21–28 (1992).
3. Brookhaven Protein Data Bank, Version of January, 1990; Bernstein, F.C., Koetzle, T.F., Williams, G.J.B., Meyer Jr., E.F., Brice, M.D., Rodgers, J.R., Kennard, O., Shimanouchi, T. and Tasumi, M., "The protein data bank: A computer-based archival file for macromolecular structures", *J. Mol. Biol.* **112**, 535–542 (1977); Abola, E.E., Bernstein, F.C., Bryant, S.H., Koetzle, T.F. and Weng, J.,

"Protein data bank", *Crystallographic Databases — Information Content, Software Systems, Scientific Applications*, eds. F.H. Allen, G. Bergerhoff and R. Sievers (Data Commission of the International Union of Crystallography, 1987) pp. 107–132.

4. TEK Command Reference Manual, Part No. 070-5141-02, Product Group 16 (1987) (Tektronix, Inc., P.O. Box 500, Beaverton, Oregon 97077).

5. Silicon Graphics Computer Systems, Document # 007-1210-020, Version 2.0 (1990) (Silicon Graphics, Inc., 2011 N. Shoreline Blvd., Mountain View, California 94039-7311).

6. Daintith, J., ed., *A Concise Dictionary of Chemistry* (Oxford University Press, 1990).

7. Foley, J. and van Dam, A., *Fundamentals of Interactive Computer Graphics* (Addison-Wesley, 1982) pp. 582–583.

8. Voet, D. and Voet, J.G., *Biochemistry* (John Wiley & Sons, 1990).

9. Walker, P.M.B., ed., *Cambridge Dictionary of Biology* (Cambridge University Press, 1989).

10. Watson, J.D., Hopkins, N.H., Roberts, J.W., Steitz, J.A. and Weiner, A.M., *Molecular Biology of the Gene* (Benjamin/Cummings Publishing Company, Inc., 1987).

Glossary

alpha helix: An element of protein structure formed by a polypeptide chain turns regularly about itself into the shape of a cylinder. The shape is stabilized by hydrogen bonding.

aromatic compound: An organic compound that contains a benzene ring in its molecules or that has chemical properties similar to benzene.

beta sheet: An element of protein structure formed by hydrogen bonding between parallel polypeptide chains.

Brookhaven protein data bank: An archival repository for the results of macromolecular structure studies such as proteins, tRNAs, polynucleotides, viruses, and carbohydrates. Two classes of information are collected, stored, and distributed: Atomic coordinates and structure factor-phase data. The center is funded by the US National Science Foundation and National Institutes of Health and located at the Brookhaven National Laboratory.

buffer: An area of main storage in a computer.

covalent bond: A chemical bond is a strong force of attraction holding atoms together in a molecule or crystal. A covalent bond is a chemical bond, formed by sharing of valence electrons (cf., ionic bond).

dihedral: An angle formed by the intersection of two planes. The *dihedral angle* is the angle formed by taking a point on the line of intersection and drawing two lines from this point, one in each plane and perpendicular to the line of intersection.

Gouraud shading: A way to improve the appearance of an approximated polygonal surface is to use a large number of smaller polygons in the approximation. An easier way is to shade or vary the color across the polygons. The latter is called Gouraud shading.

hydrophobic: Lacking affinity for water. This should be contrasted with *hydrophilic* which means having an affinity for water.

IRIS 4D: The IRIS 4D/240GTX Silicon Graphics workstation is a computer capable of displaying its own output using integrated, high-speed graphics hardware. It is widely used for scientific graphics visualizations.

residue: Many amino acids are joined by peptide bonds to form a polypeptide chain. An amino acid unit in a polypeptide is called a residue.

torsion angle: Rotation angle about the C_α–N bond (ϕ) and the C_α–C bond (ψ) of amino acid residues in a polypeptide.

Van der Waals radius: When atoms are brought together, they experience an attracting force, but very small separations, they experience a repulsive force. These forces are the van der Waals attracting and repulsive forces, respectively. At a specific distance, the attractive and repulsive forces balance each other exactly. This distance is the van der Waals radius for the atoms in question.

VAX-Tektronix 4129: The VAX-Tektronix 4129 combination is a VAX computer that displays its output on a Tektronix graphics terminal.

Z-buffering: The z-buffer is a set of integers, one associated with each pixel on the graphics screen. As each polygon, line, point, or other character is rendered, the pixel is set to the polygon's color and the z-coordinate (this is effectively the distance to the eye) is also calculated. The z value is compared to the z-buffer value for that pixel. If the z value is smaller than the value in the z-buffer, the pixel is colored and the pixel's z-buffer value is set to the new z value. Thus, at any point in the drawing, the values in the z-buffer represent the distance to the item that is currently closest to the eye.

Gene Music: Tonal Assignments of Bases and Amino Acids

Nobuo Munakata

and

Kenshi Hayashi

We proposed a tonal assignment (solmization) of four DNA bases, i.e., guanine = D(*re*), cytosine = E(*mi*), thymine = T(*sol*), and adenine = A(*la*), in a previous note.[1] To extend the assignment to amino acids, we ordered the amino acids primarily by hydropathy index with minor adjustments to preserve groupings of similar amino acids. The scale used for the bases was expanded by repeating it up and down to encompass twenty notes. The assignment in six parenthesized groups was (Arg = D1#, Lys = F1, His = G1) (Asp = A1#, Glu = C2) (Asn = D2, Gln = F2, Ser = G2, Thr = A2) (Gly = C3, Pro = D3, Ala = E3, Cys = G3) (Tyr = A3, Trp = B3, Phe = D4) (Met = E4, Leu = F4#, Val = A4, Ile = B4) (middle C = C3). To perform sequence comparison, each sequence melody of DNA or protein is assigned to a different instrument with distinct sonority, and several sequences are played polyphonically in tutti or in canon. This work is facilitated by MIDI-sequencers and synthesizers, and our preliminary explorations are sketched.

Prelude

Our approach to audio representation of genetic information is a pragmatic and light-hearted one. We do not set any fixed goal or purpose *a priori*, nor do we claim that we have found a way to make important discoveries. Instead, we suggest this as an exercise to help handle the ever-increasing amount of sequence information and to make sequences a little bit more comprehensible and familiar. This approach may stimulate high school students learning about patterns in DNA and protein sequences. In this paper, we will demonstrate the rules of our game, which require little equipment (vocal chanting) or preferably music synthesizers.

Scale of DNA Bases

Since our original proposal in *Nature* magazine,[1] we noticed or were notified of several attempts to solmizate DNA bases[2] (D.W. Deamer, personal communication). However, most of these approaches use an octave or larger intervals, and do not produce melodies which lay persons can sing, hum, or whistle while reading or typing sequences. We think our original assignment, which uses an interval of a fifth, is a useful assignment for such purposes. Having selected a musical fifth for the note interval, we chose four tones (*re, mi, sol, la*) as shown in the middle part of Fig. 1, since only the middle arrangement is a symmetrical one with regard to the intervallic relationship among the four pitches. We assigned the four DNA bases G, C, T, A to *re, mi, sol, la*, respectively, so that the more stable complementary pair (G and C) had lower pitches than the less stable one (A and T), the purines (A and G) were outside, and the pyrimidines (C and T) were inside.

Fig. 1. Tonal assignment of DNA bases.

It is interesting to note that, according to a music theoretician, Seihin Yamanouchi, these three kinds of arrangements of four tones in fifths comprise the basic modal structures of folk music in the world.[3] His theory expands upon the basic scales by combining these scales and deriving numerous exotic modes. Though we are not concerned with the details, it is interesting to note that, in his scheme, the bottom-heavy constellation (left side in Fig. 1), which is stable and optimistic, is favored in Western and Chinese folk music, while the top-heavy one (the right side), which is elegant and pessimistic, is favored in Indonesian and Okinawan music, and the symmetrical middle one, which is simple-minded and unexpressive, is favored in Mongolian, Korean, and Japanese music. The choice of the middle one seemed appropriate not only for the reason of simplicity and symmetry, but also for the fact that there seems to exist no hierarchy in the dynamic strength among four tones, which allows us to start or end at any tone without much emotional stress or hangover.

Once the assignment is memorized, it is easy to place four fingers on music keyboards to play along with the sequences. Also, in many sequence analysis software, it is possible to assign a numeric keyboard to the bases. Thus, by assigning $G = 1$, $C = 2$, $T = 3$, and $A =$ Enter, the same fingering scheme in the numeric keyboard as in music keyboards can be used. We noticed some sequence analysis software ("The DNA Inspector" for Macintosh and "PC Gene" for IBM computers) offer a menu (play DNA-song of life) to play back DNA sequences according to our assignment, though the sound from computers is not too inspiring.

Scale of Amino Acids

In the next experiment, we wanted to make tonal assignment of amino acid residues in proteins so that protein sequences could be represented by twenty-tone melodies. There are two problems in the assignment. One is how to order amino acids in a linear array, and the other is choosing an appropriate scale. As for the ordering, we tried to use the hydropathy scale, which seems to describe, at least from one point of view, the chemical characteristics of amino acid residues. Thus, we first aligned twenty amino acids according to the hydropathy index.[4] In order to preserve groupings of chemically similar amino acids, we made some fine adjustments, so that the amino acids in six groups, basic (Arg Lys His), acidic (Asp Glu), polar (Asn Gln Ser Thr), non-polar (Gly Pro Ala Cys), aromatic (Tyr Trp Phe), and hydrophobic (Met Leu Val Ile),[5] are contiguously aligned.

As for the scale, we first tried a chromatic (dodecaphonic) and a major seven-tone scale. However, both of these produced many rugged, offensive, and unnaturally-leaping intervals to our ears. We then tried a scale, which was an extension of the scale used for bases; layering it up and down in the intervals of fifths from the starting *re, mi, sol,* and *la* (D3, E3, G3, and A3). Interestingly, the scale so constructed consisted entirely of the three kinds of the scales mentioned above, combining the scales of the world. Our choice for the lowest (D2#) and the highest (B5) tone is arbitrary, but the range seemed roughly to correspond to that covered by human voices from bass to soprano. Our choice assigning the hydrophobic ones to the highest tones is also arbitrary, but, since these residues often occupy structurally critical positions and play conspicuous roles in similarity comparison, it may be rationalized to some extent. Thus, we arrived at the scale shown in Fig. 2. In practice, this scale sounded rather neutral and inoffensive in most cases, and sometimes produced surprisingly impressive melodies.

Fig. 2. Tonal assignment of amino acid residues.

Exercises

Now that the basic rules are set, there are various ways to practice. We advice you to memorize at least the base-to-tone assignment, so that you can sing along any DNA sequence. If you have a music keyboard around, you can play it at your leisure. As for amino acids, the assignment is not easily memorized, so mark the keys with the names of amino acids, and plod along to play sequences. However, this is certainly a tedious practice, and it is also impossible to perform sequence-melody comparison in this way (unless you belong to a choir or a chamber orchestra specializing in experimental music). Here, we think the help offered by the MIDI (musical instrument digital interface) is necessary.

In the mid-1980s, MIDI instruments became widely available as home-entertainment gadgets, and we started using the lowest-cost assembly comprised of a mini-keyboard, a multitimber module, and a sequencer. Today, these three elements are combined even in bottom-end keyboard synthesizers. With one of these, sequence melodies can be recorded as MIDI-data and played

Two Zippers

**for 6 percussions
arr. by N. Munakata**

Fig. 3. Part of a song titled "Two Zippers" (arrangement by N. Munakata). Regions rich in basic amino acids and leucine repeats of mouse *c-fos* and ASV *jun* sequences are played by a six-percussion ensemble. The *fos* sequence is one beat ahead of the *jun*.

Song of Courtship and Clocks

Fig. 4. Score of a song titled "Song of Courtship and Clocks" (arrangement by N. Munakata). Amino acid sequences of Drosophila *period* gene product and Neurospora *frq* gene product are played by harp and clarinet, respectively.

Fig. 4. (*Continued*)

back in eight- or sixteen-voice polyphony, and various possibilities offered by a synthesizer (including sound synthesis and sampling) can be explored. It is also advisable to use a sequencer program for a personal computer to ease experimental manipulation and file-handling.[6] For those who do not want to touch a keyboard, the sequence data in text files can be converted to MIDI-data with the help of some application programs on personal computers (see Appendix as an example). The scale of amino acids sounds often clearer and more vivid when played in a "just" intonation than in an "equal-tempered" one, so the use of synthesizers with microtuning capabilities is preferred.

A general guideline for the composition of gene music is that each sequence melody is assigned to a different instrument with distinct sonority, and several of the melodies are played polyphonically. Hence, horizontal (polyphonic) and not vertical (harmonious) listening seems essential to comprehend and appreciate the music, like listening to Renaissance music in Western tradition.

Similar sequences can be played either in phase (in tutti) so that the non-identity will be spotted easily or in canon using one- to several-beat offset. One-beat offset may be used to stress on unique repeating patterns like leucine repeats in zipper proteins (Fig. 3). In canons of several-beat offset (six to nine seems appropriate), similar melodies with variations are handed down successively to different voices like flowing waves.

We do not set any rules with regard to sonority, the choice being totally dependent on personal and synthesizer's taste. Since the capacity of today's synthesizers looks limitless, we may imagine that all the genes in a given organism or the similar genes in a variety of different organisms can ultimately be given different sounds.

Since we do not want to impose any mechanical pattern, no rule with regard to tempo or rhythm has been set. Rhythmic recognition is left as a task to the audience. For example, in a piece called "Song of Courtship and Clocks" (Fig. 4), sequences of fruit-fly *period* and fungus *frq* gene products are played by two instruments. Since these proteins seemed to be involved in biological clocks, we wondered if certain rhythmicity might arise spontaneously from the music. Indeed, some intimate dialogs of unique repeating motifs were heard when listening to this piece.

Postlude

We admit that we have explored only the most superficial aspects of gene music.[7] Many future applications could be dreamed of and need be tried so that this practice evolves to be both entertaining and illuminating. For example, the processes of information transfer may be displayed on video by audio and visual aids. Also, it might be inspiring to perceive the three-dimensional structure of a protein by sliding through the backbone with an accompanying melody and a display of a changing constellation of amino acids residues.

Envisioning the music in the electronic age, Glenn Gould wrote: "The audiences would be the artists and their life would be the arts."[8] Would all life be the arts? Metaphor of life and music abounds in many cultures. What we exemplified in this attempt was one aspect of this metaphor; both genes and music are made of linear and quantized information which underlies unfathomable diversity and mystery. Though the accumulation of sequence information is astounding, we are not confident about how to disentangle the intricate logic of life's composition. It may be useful to become a composer or an audience to elucidate it.

Acknowledgements

We thank Drs. Aldo Scarpa (Universita di Verona) and Anton Platz (Karolinska Hospital) for comments and criticism upon reading the manuscript and listening to the tape.

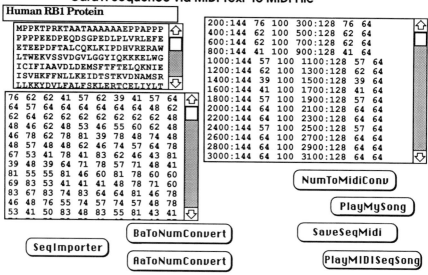

Fig. 5. "HyperCard" illustration for the conversion of sequence data to MIDI-file.

Appendix

To convert sequence data in a text-file format (e.g., from GenBank or NBRF-PIR data base) to MIDI-note files without going into MIDI-programming, a commercial program "HyperMIDI 2.0" (EarLevel Engineering) working within "HyperCard" of Macintosh computers may be used. Figure 5 is an illustrative example to perform this conversion. The text-file sequence data is imported into an upper-left field (aabaSeq) and converted to MIDI-note numbers in a lower-left field (numSeq) (script 1). This then is converted to time-stamped MIDI-note on/off data in a right-field (seqArray) (script 2) and saved as a MIDI-file (script 3).

```
(1)  –this is for amino acids; make appropriate changes in lookup for bases
     put empty into card field numSeq
     put empty into lookup
     repeat 65 times
        put "0" after lookup
     end repeat
     put "64 0 67 46 48 74 60 43 83 0 41 78 76 50 0 62 53 39 55 57 "&¬
     "0 81 71 0 69 0 " after lookup
     put hmConvert ("charsToNums", card field "aabaSeq",lookup) ¬
     into card field numSeq
(2)  put empty into card field seqArray
     put number of characters of card field aabaSeq into numRes
     set cursor to watch
     put empty into seqTemp
     repeat with i = 1 to numRes
        put 200 * i & ":144 " & word i of card field numSeq && "100 "¬
        & 200 * i + 100 & ":128 " & word i of card field numSeq && "64 "¬
        after seqTemp
     end repeat
     put seqTemp into card field seqArray
     –can be skipped to write Midifile directly
     end repeat
(3)  put card field seqArray into myMidiSeq
     get newFileName("Midi")
     if it is not empty then
        get hmMIDIfile("write",it,myMidiseq)
     end if
```

References

1. Hayashi, K. and Munakata, N., "Basically musical", *Nature* **310**, 96 (1984).
2. Ohno, S. and Ohno, M., "The all persuasive principle of repetitious recurrences governs not only coding sequence construction but also human endeavor in musical composition", *Immunogenetics* **24**, 71–78 (1986).
3. Yamanouchi, S., *Nation and the Mode* (Minzokugeinozenshu Kankokai, 1962) pp. 1–254 (in Japanese).
4. Kyte, J. and Doolittle, R.F., "A simple method for displaying the hydropathic character of a protein", *J. Mol. Biol.* **157**, 105–132 (1982).
5. Alberts, B., Bray, D., Lewis, J., Raff, M., Roberts, K. and Watson, J.D., *Molecular Biology of the Cell*, 2nd ed. (Garland Publishing, Inc., 1989) pp. 54–55.
6. Midi-sequencer and notation softwares are commercially available for all major personal computers. We used here Atari 1040ST with "Notator" (C-Lab) mainly for economical reason. A software, "Max" (Opcode Systems) for Macintosh computers, has been very helpful for the composition of gene music. As for synthesizers, we used TX81Z, TX802, and SY77 (all manufactured by Yamaha) because of the FM synthesis and microtuning capabilities.
7. From time to time, we have collected short pieces of music to see how our rules work, and the current version of the tape will be supplied to requesters who provide a blank C60 cassette tape within our capacity to handle requests. The first volume contains the following twelve pieces arranged and executed by N. M.: Code Table Song, Homeo Box Canon, Duet of AIDS, Responsorium for Major Sperm Protein Promoters, SOS Box Ricercare, Etude of a Gene, Etude of an Operon, Four Helix-Turn-Helix Degenerating, Song of Courtship and Clocks, Two Zippers, Chorale of Ras, and Four Classical Switches.
8. Gould, G., "The prospects of recording" (1966). Reprinted in Page, T., ed., *The Glenn Gould Reader* (Faber and Faber Limited, 1984) pp. 331–353.

Glossary

base: A part of each building block of DNA strand, which can either be adenine, guanine, thymine, or cytosine.

purine: A group of organic bases (adenine and guanine) found in DNA.

pyrimidine: A group of organic bases (thymine and cytosine) found in DNA.

hydropathy index: A value assigned to each amino acid with regard to relative affinity to water.

polyphonic music: A style of music consisting of independent melodies played by different voices or instruments.

tutti: All voices or instruments playing together.

***canon*:** Successive imitation of a melody by one or more voices at fixed intervals.

***dodecaphonic*:** Of twelve chromatic pitches.

***multitimbre module*:** A synthesizer unit that can produce several independent sounds simultaneously.

***just intonation*:** A tuning system consisting of acoustically pure intervals within the octave.

***sonority*:** A sound defined by timbre or the content of harmonics.

***zipper proteins*:** Proteins containing a domain characterized by a unique repeating pattern of leucine residues.

***GenBank*:** A genetic sequence data base operated by the National Center for Biotechnology Information and distributed by IntelliGenetics Inc.

***NBRF-PIR*:** A protein sequence database operated by the National Biomedical Research Foundation.

Diagrammatic Representation of Base Composition in DNA Sequences

In this paper we review and develop the "diagrammatic representation" of base composition in DNA sequences and discuss its applications. The main idea of the diagrammatic representation is that the occurrence frequencies of bases A, C, G, and T in a DNA sequence are associated with the four distances to four faces from a point within a regular tetrahedron, whose edge is equal to $\sqrt{6}/2$. This point is regarded as a representation or a mapping of the frequencies of bases. The mapping points are projected to some coordinate planes for convenience. The projection of the points constitutes a diagrammatic representation of the frequencies of bases. We have shown that the distribution of projection points for 1490 human protein coding sequences, obtained from the Gen-Bank Genetic Sequences Data Bank (Release 62.0, Dec., 1989), constitutes a particular pattern in such diagrams, which is different from those of other species. We suggest that diagrams like these may be regarded as a characteristic representation for each species. These diagrams should be very useful in the study of the molecular evolution of organisms. The application of the "tetrahedron lattices" is also discussed.

Introduction

In the last few years, the number of known DNA sequences has grown quickly. The size of the database will grow even faster in the near future. The analysis of base composition in DNA sequences plays an important role in molecular biology. Since the relevant data get larger and larger, analysis becomes more difficult. In view of this, we have developed a geometrical analysis approach, in particular a diagrammatic representation of base composition of DNA sequences. Such a diagrammatic representation should be simple and intuitive so that it helps summarize large amounts of data in a readily perceivable form. The human observer can easily draw some useful conclusions by looking at such diagrams.

A DNA sequence generally consists of four bases, A, C, G, and T. We observe that the four bases have some symmetry. There are two purines (A and G) and two pyrimidines (C and T). Also, there are two complementary base pairs, A–T and G–C. We believe that such a tetrad symmetry must have some inherent connection with the regular tetragon and the regular tetrahedron. It is possible to find the desired diagrammatic representation by using the connection between the four DNA bases and the regular tetragon as well as the regular tetrahedron.

Diagrammatic Representation

Consider an mRNA sequence or a one-strand sequence of DNA double helix with n bases. Suppose that the occurrence numbers of bases A, C, G, and T (or U) in this sequence are denoted by $n(A)$, $n(C)$, $n(G)$, and $n(T)$, respectively. Obviously, $n(A) + n(C) + n(G) + n(T) = n$. The occurrence frequencies for the bases A, C, G, and T are denoted by a, c, g, and t, respectively. They are defined as follows

$$a = n(A)/n, \quad c = n(C)/n, \quad g = n(G)/n, \quad t = n(T)/n. \tag{1}$$

According to Eq. (1) we have

$$a + c + g + t = 1, \quad 0 \leq a, c, g, t \leq 1. \tag{2}$$

Equation (2) plays a key role in this study. These equations describe the constraint conditions applied to the four real numbers, a, c, g, and t. Now consider a regular tetrahedron. There exists a simple theorem about the regular tetrahedron, which can be found in any text book on elementary geometry. This theorem may be expressed as follows.

Theorem

The sum of the four distances to the four faces from any point within a regular tetrahedron must be equal to a constant, its height.

Now imagine a special regular tetrahedron (RT), whose edge length is equal to $\sqrt{6}/2$, i.e., its height is equal to 1. Let the four RT faces be called A-face, C-face, G-face, and T-face, respectively, and let the distances of a point P within this RT to the A-, C-, G-, and T-face be equal to a, c, g, and t, respectively, then the four real numbers (a, c, g, t) can be represented by point P according to the previous theorem. Such a representation is called a "mapping" in mathematics. The four real numbers (a, c, g, t) are mapped onto a point P within the RT. In other words, if the four real numbers (a, c, g, t) are given, the position of point P within the RT can be uniquely determined. For example, if $a = c = g = t = 1/4$, the mapping point P coincides with the center of the RT. Such a mapping satisfies all of the constraint conditions listed in Eq. (2). On the other hand, if the position of point P is given, the four real numbers, a, c, g, and t, can be uniquely determined by calculating the distances of this point to the four faces of the RT. So, such a mapping is called a mapping of one-to-one correspondence.

Imagine that there are many mapping points representing a number of DNA sequences. These points are generally distributed in a three-dimensional space. However, the human observer is familiar with expressing the point distribution on a piece of paper, i.e., on a two-dimensional space. We thus project these points onto some planes. Referring to Fig. 1, consider an RT–BCGH. Let the regular triangles $\triangle$BCG, $\triangle$BGH, $\triangle$BCH, and $\triangle$CGH represent the A-, C-,

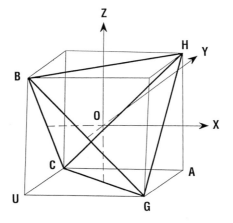

Fig. 1. A cube and its inscribed regular tetrahedron. Note that the three middle lines of the regular tetrahedron, crossing in its center, are perpendicular to each other. The coordinate system is set up by using the three middle lines.

G-, and T-face, respectively. The line connecting the middle point of an edge to the opposite edge is called the middle line of the RT. There are a total of three middle lines in an RT, crossing the center O of the RT. They are perpendicular to each other. We can set up a Cartesian coordinate system OXYZ by using the three middle lines, as shown in Fig. 1. Note that the projection of an RT onto any coordinate plane in this coordinate system is just a regular tetragon! For convenience we introduce the reduced coordinate system, Oxyz, such that X = $\sqrt{3}/4x$, Y = $\sqrt{3}/4y$, and Z = $\sqrt{3}/4z$. In such a coordinate system, the side length of the regular tetragon, the projection of the RT, is equal to 2.

Once the reduced coordinate system is set up, we want to establish the equations expressing the relation between the coordinates (x, y, z) and the four frequencies (a, c, g, t). The derivation of such equations is considerably tedious but rather elementary. We give only the final result as follows[1]

$$x = 2(a + g) - 1 = 1 - 2(c + t), \tag{3a}$$

$$y = 2(a + c) - 1 = 1 - 2(g + t), \tag{3b}$$

$$z = 2(a + t) - 1 = 1 - 2(g + c), \tag{3c}$$

or

$$\begin{pmatrix} a \\ c \\ g \\ t \end{pmatrix} = \frac{1}{4} \begin{pmatrix} 1 \\ 1 \\ 1 \\ 1 \end{pmatrix} + \frac{1}{4} \begin{pmatrix} 1 & 1 & 1 \\ -1 & 1 & -1 \\ 1 & -1 & -1 \\ -1 & -1 & 1 \end{pmatrix} \cdot \begin{pmatrix} x \\ y \\ z \end{pmatrix}. \tag{4}$$

The set of equations under Eq. (3) is called the mapping formula; while the set of equations under Eq. (4) is called the reversal mapping formula.

According to Eq. (3) we can judge the values of a, c, g, and t by observing the position of the projection point on the coordinate plane. We first discuss the x–y plane. Before continuing our discussion we should remind the reader that we usually use A, C, G, and T to represent the DNA bases, but at the same time we use the same symbols to represent the vertices of the RT or the regular tetragon. We hope this does not cause confusion. According to Eqs. (3a) and (3b), if the DNA sequence consists of only the same kind of base A, i.e., $a = 1$, $c = g = t = 0$, we find $x = 1$, $y = 1$. Referring to Fig. 2(a), the projection position of the mapping point on the x–y plane in this case is just on the vertex A of the regular tetragon. In the following, we shall use a simple symbol to denote the above circumstance, Vertex A: $a = 1$, $c = g = t = 0$. We summarize the result for the x–y coordinate plane as follows.

Vertex A: $a = 1$, $c = g = t = 0$,   Vertex G: $g = 1$, $a = c = t = 0$,

Vertex C: $c = 1$, $a = g = t = 0$,   Vertex T: $t = 1$, $a = c = g = 0$.
Side AG: $a + g = 1$, $c = t = 0$,   Side GT: $g + t = 1$, $a = c = 0$,
Side TC: $t + c = 1$, $a = g = 0$,   Side CA: $a + c = 1$, $g = t = 0$.
$x = 0 : a + g = 1/2$, $x > 0 : a + g > 1/2$, $x < 0 : c + t > 1/2$,
$y = 0 : a + c = 1/2$, $y > 0 : a + c > 1/2$, $y < 0 : g + t > 1/2$.
First quadrant: $a + g > 1/2$ and $a + c > 1/2$,
Second quadrant: $c + t > 1/2$ and $a + c > 1/2$,
Third quadrant: $c + t > 1/2$ and $g + t > 1/2$,
Fourth quadrant: $a + g > 1/2$ and $g + t > 1/2$.
Diagonal AOT: $g = c$, $\triangle$AGT: $g > c$, $\triangle$ATC: $c > g$,
Diagonal GOC: $a = t$, $\triangle$AGC: $a > t$, $\triangle$GTC: $t > a$.
$\triangle$AOG, called region I: $a > t$ and $g > c$,
$\triangle$AOC, called region II: $a > t$ and $c > g$,
$\triangle$COT, called region III: $t > a$ and $c > g$,
$\triangle$GOT, called region IV: $t > a$ and $g > c$.
Center point O: $a = c = g = t = 1/4$.

Referring to Fig. 2(b), similarly, the following is obtained for the y–z coordinate plane by using Eqs. (3b) and (3c).

Vertex A: $a = 1$, $c = g = t = 0$,   Vertex C: $c = 1$, $a = g = t = 0$,
Vertex G: $g = 1$, $a = c = t = 0$,   Vertex T: $t = 1$, $a = c = g = 0$.
Side AC: $a + c = 1$, $g = t = 0$,   Side CG: $c + g = 1$, $a = t = 0$,
Side GT: $g + t = 1$, $a = c = 0$,   Side TA: $a + t = 1$, $c = g = 0$.
$y = 0 : a + c = 1/2$, $y > 0 : a + c > 1/2$, $y < 0 : g + t > 1/2$,
$z = 0 : g + c = 1/2$, $z > 0 : g + c < 1/2$, $z < 0 : g + c > 1/2$.
First quadrant: $a + c > 1/2$ and $g + c < 1/2$,
Second quadrant: $g + t > 1/2$ and $g + c < 1/2$,
Third quadrant: $g + t > 1/2$ and $g + c > 1/2$,
Fourth quadrant: $a + c > 1/2$ and $g + c > 1/2$.
Diagonal AOG: $c = t$, $\triangle$ACG: $c > t$, $\triangle$AGT: $t > c$,
Diagonal COT: $a = g$, $\triangle$ACT: $a > g$, $\triangle$CGT: $g > a$.
$\triangle$AOC, called region I: $a > g$ and $c > t$,
$\triangle$AOT, called region II: $a > g$ and $t > c$,
$\triangle$TOG, called region III: $g > a$ and $t > c$,
$\triangle$COG, called region IV: $g > a$ and $c > t$.
Center point O: $a = c = g = t = 1/4$.

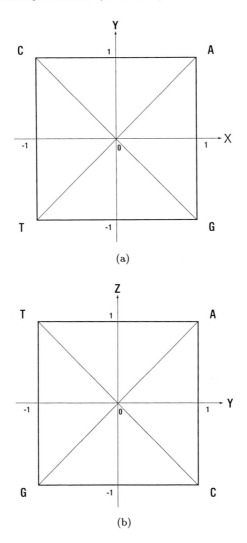

Fig. 2. Projection of the whole regular tetrahedron onto (a) the x–y coordinate plane and (b) the y–z plane. Note that each projection is a regular tetragon. For convenience, the vertices of each tetragon (A-vertex, C-vertex, G-vertex, and T-vertex) are represented by A, C, G, and T, respectively.

The reader may perform similar analysis on the x–z coordinate plane by using Eqs. (3a) and (3c). However, since the distribution of points in a three-dimensional space can be completely described by the projection onto at least two coordinate planes, we shall hereafter use only the x–y and y–z planes.

Suppose that there are two sequences whose base compositions are denoted by (a_1, c_1, g_1, t_1) and (a_2, c_2, g_2, t_2), respectively. Their mapping points in the RT are P_1 and P_2, respectively. Then the distance between P_1 and P_2 is calculated as[1]

$$d_{12} = \frac{\sqrt{3}}{2} \left[(a_1 - a_2)^2 + (c_1 - c_2)^2 + (g_1 - g_2)^2 + (t_1 - t_2)^2 \right]^{\frac{1}{2}}. \qquad (5)$$

We define d_{12} as the "distance" between two DNA sequences.

In summary, the occurrence frequencies of bases in a DNA sequence defined in Eq. (1) are mapped to a point within the RT. This point is then projected onto the x–y and y–z coordinate planes, respectively, according to the Eq. (3) for further study. Generally, we deal with a number of points. We can draw some conclusions by observing the distribution of these points in the two projection planes at the same time. The distance between any two mapping points calculated by Eq. (5) may be used to measure the difference between the two corresponding sequences.

Application and Discussion

The occurrence frequencies of bases in the first, second, and third codon position for the coding sequences of ninety species were calculated and these data were studied by the diagrammatic representation method described here.[2] The ninety species include different animals such as human, hamster, mouse, rat, bovine, dog, pig, rabbit, sheep, and chicken, as well as virus, phage, bacteria, etc. The name of each species is listed in Table 2 of reference 3. In this paper, we give another example of the application of our method. The codon usages for 1490 human proteins were analyzed by using the nucleotide sequence data obtained from the GenBank Genetic Sequence Data Bank (release 62.0, December, 1989).[3] The names of 1490 human proteins are listed in Table 3 of reference 3. Based on these data we have calculated the occurrence frequencies of four bases for each of the 1490 human proteins. These frequencies are mapped to an RT and then projected onto the x–y plane, as shown in Fig. 3(a), and the y–z plane, as shown in Fig. 3(b). In each of these figures, there are 1490 points representing 1490 human protein coding sequences, respectively. Refer-

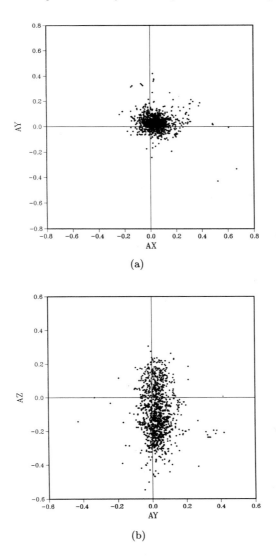

Fig. 3. The occurrence frequencies of bases A, G, G, and T (or U) in each of 1490 human protein coding sequences are mapped to the interior of the regular tetrahedron and then projected to (a) the x–y coordinate plane and (b) the y–z plane. The detailed names of these human proteins were listed in Table 3 of reference 3. Note that in each diagram there are a total of 1490 points, representing 1490 human protein coding sequences, respectively. These frequencies are obtained by averaging over those in the first, second and third codon positions. The character "A" in "AX", "AY", and "AZ" in the figures denoted this fact.

ring to Fig. 2(a), as we see from Fig. 3(a), most of the projecting points are distributed at the region of $\triangle AGC$. This means that $a > t$. Furthermore, since most of the points are in the region of $x > 0$, this implies that $a + g > 1/2$ or the purine bases are dominant for most of these human protein coding sequences. Now turn to Fig. 3(b) and refer to Fig. 2(b). About one-third of the points are in the region of $z > 0$, i.e., $g + c < 1/2$; and two-thirds are in the region of $z < 0$, i.e., $g + c > 1/2$. The overall average of $g + c$ seems to be greater than $1/2$. Furthermore, nearly all of the points are situated at the region of $y \simeq 0$ and are roughly symmetrical with respect to the z-axis ($y = 0$). This means that $a + c \simeq 1/2$. Summarizing the results seen in Figs. 3(a) and (b), we find that the mapping points are distributed in a column-like region in a three-dimensional space, and $a + g > 1/2$, $a + c \simeq 1/2$, and $a > t$, $g > c$ for most of 1490 human protein coding sequences.

All of the above peculiarities about the distribution of the mapping points for 1490 human protein coding sequences constitute a particular pattern. We have found that the distribution of patterns for different species are different. The work is still in progress and the details will be discussed elsewhere. More than ten years ago, Grantham pointed out that the mRNA sequences contain information other than that spelling for encoding proteins.[4] This information is contained mainly in the degenerate bases of the third codon position. It is well known that the choice of bases in the third codon position is species-specific. This fact was referred to as a "genome hypothesis" by Grantham,[4] or a "codon dialect" by Ikemura.[5] Recently, we have found that the distribution patterns in the diagrammatic representation for the base composition in the third codon position are strongly species-specific. This further confirms the observation of Grantham and Ikemura. We suggest that the diagrams expressing the base composition in the third codon position of the coding sequences for each species may be regarded as a characteristic representation for that species. The situation is somewhat similar to the characteristic light spectrum pattern for each atom in physics. Therefore, it is important to compare different species by comparing their corresponding distribution patterns of the characteristic diagrams. This would be very useful in the study of the classification and the evolution of organisms.

Another possible application relates to the classification of proteins based on the "distance principle". Imagine that there are three points P_1, P_2, and P_3 within the RT, representing protein-1, -2, -3, respectively. The distances between these points can be calculated by Eq. (5). Suppose that $d_{12} \ll d_{13}$.

Then it is reasonable to think that protein-1 is more similar to protein-2 than to protein-3. Therefore, the proteins may be classified according to the positions of their mapping points in the RT. Techniques to deal with such a classification are available, e.g., the technique of cluster analysis.[6]

Let us discuss the applications of a "tetrahedral lattice". A tetrahedral lattice is defined as a set of all possible mapping points in the RT for all the DNA sequences which have the same length ($n = 1, 2, 3, 4, 5, 6, \ldots$). First, we discuss the extreme case: $n = 1$, i.e., the sequence consists of only one base. Clearly, there are only four possibilities: Either A or C, or G or T. So the tetrahedral lattice (TL) in this case consists of only four points, the four vertices of the RT. For example, for the "sequence" A, we have $a = 1$, $c = g = t = 0$, and its mapping point is just the A-vertex of the RT. The A-vertex is one of the vertices of the RT, whose distance to the A-face is equal to 1. We use similar arguments for the C-, G-, and T-vertex. In our coordinate system, the lines OA-, OC-, OG-, and OT-vertex constitute four real vectors. A realistic DNA sequence may be regarded as a vector sequence consisting of these four real vectors. The Fourier transform of this vector sequence has been used to study the periodicity of the realistic DNA sequences.[7] Furthermore, Pickover has used the four vectors to display the DNA sequences.[8] The DNA sequence is displayed in three-dimensional space by connecting the "head" of one vector to the "tail" of the next vector for each base.[8] Now turn to the case of $n = 2$. There are a total of $4^2 = 16$ possible sequences. They are AA, CC, GG, TT, AC, AC, $\ldots$, . However, some sequences, e.g., AC and CA, are "degenerate" in the sense that they correspond to the same mapping point. So there is a total of 10 lattice points in this case, in which four are on the four vertices, six on the middle points of six edges of the RT. The case of $n = 3$ seems to be more interesting. In this case there is a total of $4^3 = 64$ possible sequences, or 64 codons. By considering the case of "degeneracy", the total number of lattice points is just twenty, the same number of amino acids. Among the twenty lattice points four are on the vertices, twelve on the edges, and the remaining four on the centers of four faces of the RT. Such a tetrahedral lattice was first proposed by Trainor *et al.*, in 1984.[9] They have found the relation between the twenty TL points with the twenty amino acids. Their study has led to an interesting speculation on the origin and development of the genetic code.[9] When $n \geq 4$, the TL points begin to enter into the interior of the RT. When n increases, the number of TL points increases quickly. In all cases, whatever the value of n, the TL points are distributed symmetrically with respect to the center O of the RT. The tetrahedral lattice transforms the symmetry of DNA sequences to the symmetry of space of TL points. We think that the concept of tetrahedral lattices is worthy of further study.

Conclusion

The occurrence frequencies of bases in a DNA sequence are mapped to points within a regular tetrahedron, whose edge length is $\sqrt{6}/2$. The mapping points are then projected onto some coordinate planes. The projection points are always situated within a regular tetragon in this coordinate system. A particular distribution of the mapping points or their projection points corresponds to a particular set of DNA sequences. We have found that the distribution patterns corresponding to the protein coding sequences are species-dependent. In particular, the distribution patterns in the diagrams corresponding to the occurrence frequencies of four bases in the third codon position of the coding sequences are strongly species-specific. We suggest that the diagrams expressing the distribution pattern for one species may be regarded as a characteristic representation for this species. This would be very useful in the study of the classification and the evolution of organisms. The diagrammatic technique can be used for both coding or non-coding DNA sequences.

The concept of the tetrahedral lattice has been proposed in this paper. A tetrahedral lattice is defined as a set of all possible mapping points wthin the RT for all the DNA sequences which have the same length. The tetrahedral lattice transforms the symmetry of the DNA sequences to the symmetry of space of the lattice points. We have discussed some applications of such lattices. Further study of the concept seems warranted.

The diagrammatic representation presented is an intuitive tool for the study of DNA sequences. Its main merit is the help it provides researchers in summarizing massive amounts of data in a readily perceivable form. It is also simple and easy to use. Once the occurrence frequencies of bases are given, one can easily draw the point distribution diagram by using Eq. (3). We expect that this technique will become a conventional tool for the relevant reseachers. Recent development of the technique can be found in references 10–14.

Acknowledgement

Invaluable and stimulating discussions with Ren Zhang are gratefully acknowledged. The present study was supported in part by the grant #19325002 and #19577104 from the China Natural Science Foundation.

References

1. Zhang, C.T. and Zhang, R., "Diagrammatic representation of the distribution of DNA bases and its applications", *Int. J. Biol. Macromol.* **13**, 45–49 (1991).
2. Zhang, C.T. and Zhang, R., "Analysis of distribution of bases in the coding sequences by a diagrammatic technique", *Nucl. Acids. Res.* **19**, 6313–6317 (1991).
3. Wata, K., Aota, S., Tsuchiya, R., Ishibashi, F., Gojobori, T. and Ikemura, T., "Codon usage tabulated from the GenBank genetic sequence data", *Nucl. Acids. Res.* **18**, r2367–r2411 (1990).
4. Grantham, R., "Workings of the genetic code", *Trends Biochem. Sci.* **5**, 327–330 (1980).
5. Ikemura, R., "Codon usage and tRNA content in unicellular and multicellular organisms", *Mol. Biol. Evol.* **2**, 13–34 (1985).
6. Anderberg, M.R., *Cluster Analysis for Applications* (Academic Press, 1973).
7. Silverman, B.D and Linsker, R., "A measure of DNA periodicity", *J. Theor. Biol.* **118**, 295–300 (1986).
8. Pickover, C.A., "DNA and protein tetragram: Biological sequences in tetrahedral movements", *J. Molec. Graphics* **10**, 2–6 (1992).
9. Trainor, L.E.H., Rowe, G.W. and Szabo, V.L., "A tetrahedral representation of poly-codon sequences and a possible origin of codon degeneracy", *J. Theor. Biol.* **108**, 459–469 (1984).
10. Chou, K.C. and Zhang, C.T., "Diagrammatization of codon usage in 339 human immunodeficiency virus proteins and its biological implication", *AIDS Res. and Human Retroviruses* **8**, 1967–1976 (1992).
11. Zhang, C.T. and Zhan, Y., "Analysis on the distribution of bases for 1487 human protein coding sequences", *J. Theor. Biol.* **167**, 161–165 (1994).
12. Zhang, C.T. and Chou, K.C., "Graphic analysis of codon usage strategy in 1490 human proteins", *J. Protein Chem.* **12**, 329–335 (1994).
13. Zhang, R. and Zhang, C.T., "Z curves, an intuitive tool for visualizing and analyzing the DNA sequences", *J. Biomol. Struc. Dynamics* **11**, 767–782 (1994).
14. Zhang, C.T. and Chou, K.C., "A graphic approach to analyzing codon usage in 1562 *E. Coli* protein coding sequences", *J. Mol. Biol.* **238**, 1–8 (1994).

A Transforming Function for the Generation of Fractal Functions from Nucleotide Sequences

José Campione-Piccardo

Nucleic sequences are conveniently represented as arrays of four symbols corresponding to the four nucleotides most prevalent in nature. This representation is essentially qualitative and not easily amenable to mathematical analysis. For several reasons, some of which are discussed below, it is of interest to represent nucleotide sequences as quantitative functions. A quantitative mathematical representation of such arrays has to take into account the qualitative, discrete essence of nucleotide sequences. The method proposed in this report is based on the assessment of the coordinates for the center of mass of a 3-D mechanical analog. Each of the four vertices in a tetrahedron is assigned one of the four bases in a nucleotide sequence. The entire mass of the system is assumed to be applied only in these vertices, and the mass of every new base is assumed to be proportional to the entire mass already in the system. Then, for every nucleotide position, it is possible to compute the coordinates of the center of mass. The relationship between the angular coordinates of this center and the base position corresponds to a function with fractal properties.

Introduction

Graphic representation can be considered as a branch of descriptive statistics, the main role of which is to summarize data and to aid human interpretation through simplification of data and highlighting special features. Although, in some special cases, as many as six different variables have been effectively represented on a 2-D surface,[6] the intrinsic limitations in the way human beings appreciate reality usually limits the number of variables to not more than three, preferably two. Since there is an interest in locating the position of special features, one of these dimensions has to represent the position within the sequence. This means that to be effective, a 2-D representation of a nucleotide sequence requires the plotting of a single variable depending on the nucleotide sequence versus the position in the sequence. The problem is thus reduced to obtaining a single variable that would best represent the composition of the sequence in terms of its four bases and their sequential position in the sequence.

The best solutions for this problem make use of Fourier transforms.[8,11,17] This permits the analysis of DNA sequences with the same arsenal of methods developed for other waveforms as, for example, the analysis of speech. Other more empirical methods have also been proposed. Vectorgrams[10,11] are derived by assigning 0 or 1 to each base, and the eight binary triplets formed by three contiguous bases are then assigned one of the eight possible cross or diagonal movements in a 2-D space. Tetragrams[12] use a similar idea in a 3-D space. In this case the problem of assigning four orientations in a 3-D space is solved by assigning the directions for each base to one of the four vertices of a tetrahedron. H-curves have also been suggested.[2] These curves are constructed by assigning each base to one of four perpendicular directions. The third dimension is assigned to the position in the sequence. For practical purposes the representation is usually reduced to a lateral or vertical projection of these curves. Some of these methods require that arbitrary values be assigned to each of the four bases. Another more pervasive problem is the difficulty in the simultaneous unbiased representation of five variables (the four bases and their position in the sequence).

Here, we explore another approach designed to transform the nucleotide sequence into a single variable function with fractional Brownian motion or brown noise behavior. This transformation is of interest due to the fact that fractional Brownian motion represents the paradigm of random fractals[15] and it would thus permit the analysis of nucleotide sequences with the same mathematical tools being developed for random fractals.

Fractals, Noise and Fractional Brownian Motion

Objects in the natural world are usually analyzed through mathematical models used to represent them. Regular-shaped objects can be analyzed using models of traditional Euclidian shapes. Common natural shapes such as coastal and mountain lines, clouds, flakes, and trees cannot be described in these terms. They are better described with recursive mathematical models creating shapes with self-similarity. This self-similarity makes these shapes appear similar no matter what scale is used for their analysis and resolution. Shapes with these characteristics are called fractals.

Relatively simple, recursive mathematical expressions can create complex deterministic fractal curves. Natural-occurring shapes with some degree of self-similarity are usually better described using random fractals generated by stochastic processes.

The graphic representation of a random variable as a function of a single variable (usually time) is called white noise. The waveform created in this way is best analyzed by the spectral synthesis method, also known as the Fourier filtering method, based on Fourier transforms of the original function.[13,14] This analysis permits the calculation of β, the spectral exponent. Random waveforms with low integer values for this coefficient are identified as noises of specific color. Thus, white, pink, brown and black noises have values of $\beta = 0, 1, 2$, and 3. As the absolute value of the spectral exponent increases, the waveform is more persistent in the sense that its graph tends to cross the middle of its vertical range less often.

A function is a fractional Brownian motion function, $V_{(x)}$, if it is a single, valued function of one variable, x (usually time); its increments, $V_{(x[i+1])} - V_{(x[i])}$, have a Gaussian distribution; and their mean quadratic difference of the increments is proportional to the increment in the independent variable:

$$\left(V_{(x[i+1])} - V_{(x[i])}\right)^2 = k \cdot \left(x_{[i+1]} - x_{[i]}\right)^{2H}.$$

It can be shown that for $H = 0.5$ the function represents Brownian motion in one dimension. For $H = 0$, the function behaves as white noise and for $H = 1$ the function corresponds to brown noise. Parameter k is a proportionality constant. Parameter H, also known as the Hurst exponent,[3,15] is a convenient measure of the persistence of a statistical phenomenon. These two constants correspond to the intercept and slope of the linear relation between the logarithmic form of the above equation:

$$\ln[(V_{(x[i+1])}) - V_{(x[i])})] = \ln(k)^2 + 2H \cdot \ln(x_{[i+1]} - x_{[i]}).$$

It can be shown that the H parameter has a simple relation with the spectral exponent, β, of the power spectrum of random noises

$$\beta = 2H + 1$$

and that H is related to the Euclidean embedding dimension, E, and the fractal dimension, D, of a fractional Brownian motion by the following equation:

$$D = E + 1 - H.$$

For comprehensive presentations on fractals and other functions described in this section, the reader is referred to references 1, 5, 7, and 15.

The barogram model

The approach in this chapter is based on the assessment of the coordinates for the center of mass of a 3-D mechanical analog. Each of the four vertices in a tetrahedron are assigned to one of the four bases in a nucleotide sequence. The entire mass of the system is assumed to be applied only to these vertices, and the mass of every new base is assumed to be proportional to the entire mass of the system. Then, the coordinates of the center of mass are computed from the moments of the system with respect to each of the coordinate axes. The variation of these coordinates with the position in the sequence represent single, variable wave forms directly related to the overall and local composition of the sequence. Since the center of mass is also known as the "barycenter" from the Greek "baros" meaning "weight", the name of "barograms" is proposed here to designate the graphic representations of the variations of these coordinates.

Barograms depend not only on the composition of the sequence, but also on the position of the tetrahedron with respect to the three coordinate axes. The arrangement used in this study can be represented by the matrix depicted in Fig. 1. Planar projections of the tetrahedron oriented according to this matrix are presented in Fig. 2.

	x	y	z
A	1	1	1
T	-1	-1	1
G	-1	1	-1
C	1	-1	-1

Fig. 1. Matrix indicating the coordinates for each of the four vertices of the tetrahedron in a 3-D space.

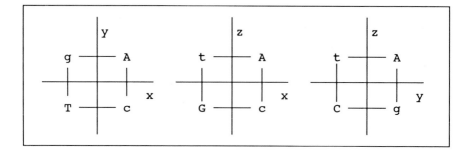

Fig. 2. Planar projections of the tetrahedral arrangements for the four different bases. The third axis in each case is positive towards the viewer. The capital letters represent vertices closer to the viewer than those represented in lower-case letters.

If $w[i]$ is the mass applied to each i point, and the $x[i]$, $y[i]$, and $z[i]$ are the coordinates of these points, then, after considering n points, the moments of the system are

$$Mx[n] = \sum_{i=1}^{n} (w[i] * x[i])$$

$$My[n] = \sum_{i=1}^{n} (w[i] * y[i])$$

$$Mz[n] = \sum_{i=1}^{n} (w[i] * z[i])$$

and the mass of the system is

$$W[n] = \sum_{i=1}^{n} (w[i]) \, .$$

The coordinates of the center of mass of the system after considering n points are

$$X[n] = Mx[n]/W[n]$$

$$Y[n] = My[n]/W[n]$$

$$Z[n] = Mz[n]/W[n] \, .$$

The plotting of these coordinates as a function of n originates waveforms related to the sequence composition. These waveforms are those referred to in this communication as "barograms".

One way to obtain a 2-D representation of these data is to consider the sphere, containing on its surface the four vertices of the nucleotide tetrahedron, and to determine the two angular coordinates of the points in which the line containing the center of mass intersects with the sphere surface. These angles can be obtained from the tangents of the angles formed by the vector of the center of mass in x–y plane and in the plane defined by the z axis and the vector itself. The plotting of these two angles as a function of the nucleotide position is referred to as "angular barograms".

The tangent of the angle in x–y plane is given by

$$t[x - y, \, n] = Y[n]/X[n] = My[n]/Mx[n] \,.$$

Similarly the tangent of the angle in the plane formed by the vector passing through the center of mass (v) and the third axis, z, is given by

$$t[z - v, \, n] = \overline{t(((Mx[n])^2 + (My[n])^2)/Mz[n])} \,.$$

These two tangents can then be used to locate the axis containing the center of mass of the system and its extrapolation to the surface of the sphere passing through the four vertices of the tetrahedron. Similar equations can be determined for the other axis arrangements. Although different angular barograms will be obtained, they will correspond to different arrangements of the same data.

If an additional point with mass $w[n + 1]$ is considered and $mx[n + 1]$ and $my[n + 1]$ are the moments of the new point respective to both axes:

$$mx[n + 1] = w[n + 1] * x[n + 1]$$

$$my[n + 1] = w[n + 1] * y[n + 1]$$

$$mz[n + 1] = w[n + 1] * z[n + 1]$$

then,

$$X[n + 1] = (Mx[n] + mx[n + 1])/(W[n] + w[n + 1])$$

$$Y[n + 1] = (My[n] + my[n + 1])/(W[n] + w[n + 1])$$

$$Z[n + 1] = (Mz[n] + mz[n + 1])/(W[n] + w[n + 1])$$

$$t[x - y, n + 1] = (My[n] + my[n + 1])/(Mx[n] + mx[n + 1])$$

which can also be represented as a partially recursive equation:

$$t[x - y, n + 1] = (t[x - y, n] + (my[n + 1]/Mx[n]))/(1 + (mx[n + 1]/Mx[n])) .$$

If the mass of each new point is made to be independent from the position of the base in the sequence, the impact of new bases on the coordinates of the center of mass will be reduced as the total mass added to the system grows. To avoid this, the mass represented by each new base can be made proportional to the total mass already in the system:

$$w[n + 1] = 1/w * W[n] .$$

Then,

$$t[x - y, n + 1] = (Y[n] + 1/w * y[n + 1])/(X[n] + k * x[n + 1]) .$$

In this manner the average amplitude of the oscillations is kept constant for the entire sequence and the features are reproduced independently of the previous composition (i.e., the same sequence in tandem reproduces the same features). Also, the proportionality constant, k, allows control of the amplitude of the oscillations. This is of interest because, depending on the value assigned to this parameter, the barogram approaches the characteristics of different kinds of noise.

The use of an angular value to represent fractals has the disadvantage of corresponding to real values enclosed within the half-open interval $0° >=$ ang $< 360°$. Since this limitation affects the calculations of fractal dimensions, the problem can be solved by keeping track of every full turn around the center. In the algorithm this can be done by adjusting the number of turns according to the variation in the angle with respect to every previous estimate. The values can then be assumed to be unbound at both sides of the interval.

The overall position of the barogram is affected by the proportion of each base in the DNA. This effect can be neutralized by dividing the coordinates assigned to each of the four bases by their proportion in the sequence. This procedure has the effect of deforming the regular icosahedron into an irregular one with vertices whose coordinates are inversely related to the abundance of the bases they represent. This allows a more even distribution of the points in the barogram and permits better evaluation of local features.

Although not explored in this chapter, the same approach could have been followed by computing the radius of gyration of the system, Rg, for every subsequent new base in the sequence:

$$Rg[n] = \left(\sum_{i=1}^{n} (x[i] - X[i])^2 \right) \Big/ n \, .$$

This value represents the mean quadratic distance from the center of mass in a given axis. It reflects the cohesiveness of the system and has been used to represent the motions of breathing proteins.[8,11]

A Pascal Implementation

In the Turbo Pascal implementation that follows, xa, xt, xg, and xc represent the tetrad arrangement of the four bases with respect to the x axis, and similarly, ya, yt, yg, yc represent the same arrangement with respect to the y axis; ta, tt, tg, and tc are the total "masses" assigned to each base; mx and my are the total moments of the system with respect to the x and y axis; i is the position of each base and seqlen is the total length of the sequence; w is the barogram weight factor. It corresponds to the inverse of the fraction of the total mass representing the increment (d) at each base position; t is the tangent of the angle of the vector for the "center of mass", and "ang" is the angle (in degrees) of this vector in the x–y plane; $b[i]$ is the array containing each of the bases of the sequence. The conversion of radians to degrees is not a requirement of the model.

```
{Procedure BaroGram;
Copyright 1991 by J. Campione.}

{Note: Declarations of types and variables and code for utility
routines are not included.}


begin

  {Read sequence from file into b array.
   - b is the array containing the sequence.
```

```
 - seqlen is the total number of bases.}
ReadSeq(b,seqlen);

{Initialize screen graph mode.}
InitialGraph;

{Initialize barogram weight factor.}
w:= 500;

{Initialize base proportions in entire sequence
(these are the base proportions in the HPV16 sequence).}
as:= 0.329;
ts:= 0.306;
gs:= 0.191;
cs:= 0.174;

{Initialize three coordinates
for each tetrahedron vertex.}
xa:=  1; ya:=  1; za:=  1;
xt:= -1; yt:= -1; zt:=  1;
xg:= -1; yg:=  1; zg:= -1;
xc:=  1; yc:= -1; zc:= -1;

{Initialize totals for each base.}
ta:= 0; tt:= 0; tg:= 0; tc:= 0;

{Initialize previous values of angles.}
oldang:= 0; oldong:= 0;

{Initialize previous values of coordinates.}
oldyy:= 0; oldxx:= 0; oldzz:= 0;

{Correct values of coordinates for the proportion
of each base in the DNA sequence.}
xa:= xa * as; ya:= ya * as; za:= za * as;
xt:= xt * ts; yt:= yt * ts; zt:= zt * ts;
xg:= xg * gs; yg:= yg * gs; zg:= zg * gs;
xc:= xc * cs; yc:= yc * cs; zc:= zc * cs;
{Initialize the number of full turns in first barogram.}
turn:= 0;

{Start loop for each position in the DNA sequence.}
for i:= 1 to seqlen do begin
```

```
{Save previous value in screen x axis.}
oldxx:= xx;

{Determine new value in screen x axis.}
xx:= (i/Seqlen) * GetmaxX;

{Divide total mass by weight factor to
obtain size for subsequent increment.}
d:= (ta+tt+tg+tc+1)/w;

{Modify coordinates depending on base at
position being considered.}
case b[i] of
'A' : ta:= ta + d;
'T' : tt:= tt + d;
'G' : tg:= tg + d;
'C' : tc:= tc + d;
end;

{Determine the momentum for the center of mass
in each of the 3 dimensions.}
mx:= ta*xa + tt*xt + tg*xg + tc*xc;
my:= ta*ya + tt*yt + tg*yg + tc*yc;
mz:= ta*za + tt*zt + tg*zg + tc*zc;


{Determine the new 3-D Cartesian
coordinates for the center of mass.}
x:= mx/(ta+tt+tg+tc);
y:= my/(ta+tt+tg+tc);
z:= mz/(ta+tt+tg+tc);

{Determine the actual size of the vector
(required to draw the second barogram).}
v:= sqrt(sqr(x) + sqr(y));

{Determine in degrees the angle of the vector on the x-y plane
(skip the value if a denominator could be zero).}
if (x <> 0) and (z <> 0) then begin
oldang:= ang;
if (x >= 0) and (y >= 0) then ang:= arctan(y/x) * 180/pi;
if (x <  0) and (y >= 0) then ang:= arctan(y/x) * 180/pi + 180;
if (x >= 0) and (y <  0) then ang:= arctan(y/x) * 180/pi + 360;
if (x <  0) and (y <  0) then ang:= arctan(y/x) * 180/pi + 180;
```

```
oldyy:= yy;

{Determine the screen y axis for the first barogram.}
yy:= getmaxy - ((ang/360) * (getmaxy));

{Correct for number of turns.}
if (oldang - ang > 180) then inc(turn,1);
if (ang - oldang > 180) then dec(turn,1);
ang:= ang + 360 * turn;

{initialize value if first base.}
if i = 1 then lastang:= ang;

{Draw the x-y barogram function on the screen.}
DrawXY(oldxx,xx,oldyy,yy);

{Determine in degrees the angle of the vector in the plane
determined by the z axis and the vector itself.}
oldong:= ong;
z:= -z;
if (z >= 0) and (v >= 0) then ong:= arctan(v/z) * 180/pi;
if (z <  0) and (v >= 0) then ong:= arctan(v/z) * 180/pi + 180;

oldzz:= zz;

{Determine the screen y axis for the second barogram.}
{Note: The angle of the first barogram moves in a 0-360 interval
while the angle of the second barogram moves in the 0-180 range}
zz:= getmaxy - ((ong/180) * (getmaxy));

{Draw the v-z barogram on the screen.}
DrawVZ(oldxx,xx,oldzz,zz);
  end;
 end;
end.
```

Angular Barograms as Fractals

The following examples and analysis make use of the nucleotide sequence of the human papillomavirus type 16 (HPV16). HPV16 is the human papillomavirus most often associated with human genital malignancies.[4] Its genome is

composed of two circular strands containing 7905 bases each.[16] Only one of
the strands codes for all the genes expressed by the virus. This strand contains
seven early open reading frames and two late ones (see Fig. 3(c)). The base
composition of this strand is: A: 32.9%, C: 17.4%, G: 19.1%, and T: 30.6%.

Figure 3 depicts angular barograms obtained from the sequence of the coding strand of HPV16 using $w = 500$. The vertical (Fig. 3(a)) and horizontal
(Fig. 3(b)) angles of the vectors passing through the center of mass are plotted
against the genome position. The open reading frames present in the coding
strand of HPV16, are depicted at the bottom of the picture (Fig. 3(c)). The
same data are presented in Fig. 4, where the vertical angle between the vector
passing through the center of mass and the y–δ axis is plotted against the angular coordinate of the same vector in the $[x$–$y]$ plane. In this representation,
each dot corresponds to a nucleotide position. The difference in nucleotide
composition between the early (left cluster) and late (right cluster) portions of
the genome are clearly noticeable.

The waveform generated by the barogram transformation bears a close
resemblance to a fractal curve, especially to fractional Brownian motion. The
following analysis provides more objective evidence for this resemblance.

As indicated above, fractional Brownian motion is generated by a variable
with Gaussian distribution and with mean quadratic increments proportional
to the variation in the independent variable (in this case the distance between
base position in the DNA sequence).

The histogram obtained for the increments in Fig. 3(b) for the HPV16 genomic sequence is presented in Fig. 5. Their distribution approaches better a
lognormal distribution than a Gaussian distribution. Interestingly, the skew-
ness of this distribution is not completely the consequence of non-randomness
due to the biological nature of the sequence. This is demonstrated by the fact
that a random sequence of the same length and similar base composition also
generates the same kind of distribution. Furthermore, a completely random
sequence in which each base has the same proportion, also generates a skewed
distribution.

The "fractalness" or self-similar nature of angular barograms can be as-
sessed by the determination of the relation between the mean quadratic deviation of the increments and the distance between the base positions being
considered. As described above for fractional Brownian motion, if angular
barograms are fractal curves the logarithm of the quadratic mean of the angular increments should be proportional to the logarithm of the distance between
the base positions used to measure the increment. The results shown in Fig. 6

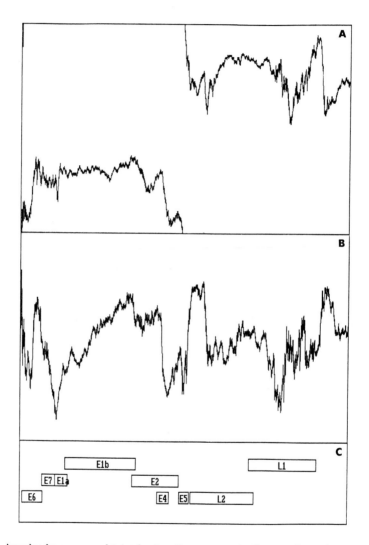

Fig. 3. Angular barograms obtained using the program in the text from the complete sequence of the coding DNA strand in the genome of papillomavirus type 16 (HPV16). In Sec. A, the ordinate indicates the angular variation in the x–y axis (as defined in Figs. 1 and 2). In Sec. B, the ordinate represents the angular variation in the plane defined by the z-axis and the vector passing through the center of mass. In both sections the abscissa represents the position in the linearized DNA (7905 bases). The left-most position corresponds to position 1 in the sequence published by Seedorf *et al.* (1985).

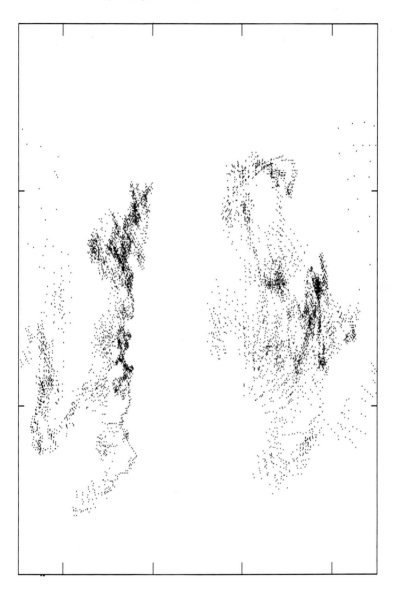

Fig. 4. 2-D representation obtained from both angular barograms in Fig. 3. The abscissa corresponds to the angular barogram in Fig. 3(a) and the ordinate corresponds to the function in Fig. 3(b). Figures 3 and 4 were directly plotted by the Pascal program on a VGA screen before being directed to an HP-Laserjet Series II printer.

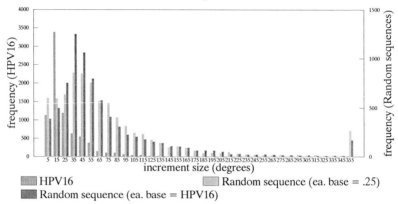

Fig. 5. Distributions of angular increments for different angular barograms. The HPV16 barogram corresponds to the same one depicted in Fig. 3(a). The random sequences were generated with the same proportion (0.25) for each of the four bases, or with the same base composition as the coding strand of HPV16.

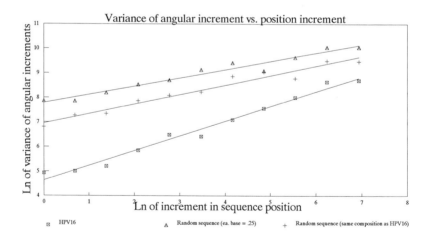

Fig. 6. Variance of angular increment versus distance in base position. The plots were obtained from the same data shown in Fig. 5. An iterative form of the Pascal code presented in the text was used to generate variance data for base distances of 1, 2, 4, 8, 16, 32, 64, 128, 256, 512, and 1024 bases. The linearity of the power relation between the variance of function increments and the distance of the positions in the DNA is an indication of self-similarity ("fractalness") in angular barograms.

indicate that this is the case. Using the fractional Brownian motion model, the slopes in those curves correspond to $2H$. The values of H and of the corresponding fractal dimensions for these angular barograms are indicated in Table 1.

It can be observed that angular barograms from natural sequences have a higher degree of persistence (as revealed by the higher value of H) than equivalent random sequences. This higher persistence is not the result of the base composition (as revealed by the fact that random sequences with the same composition as the natural sequence still have a lower value of H). Therefore,

Table 1. Values of Hurst exponent, H, fractal dimension, D, and spectral exponent, β, for the barogram data in Fig. 5. (Barogram weight factor: $w = 500$).

	H	D	β
HPV16 sequence	0.30	1.70	1.60
Random sequence (ea. base = .25)	0.17	1.83	1.34
Random sequence (same composition as HPV16 sequence)	0.19	1.81	1.38

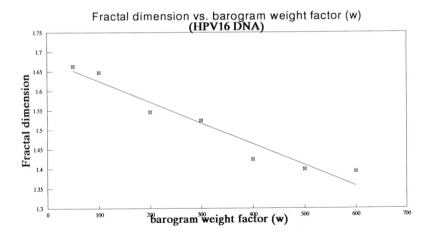

Fig. 7. Linear relation between the fractal dimension of the angular barogram and the barogram weight factor, w, for angular barogram data generated by the coding sequence of human papillomavirus type 16 (HPV16).

it is proposed that the H value for a barogram from natural sequences could be related to the non-randomness of natural genetic sequences.

It was also observed that, within a certain range, the fractal dimension of a barogram is inversely proportional to the barogram weight factor (w) (Fig. 7).

Conclusion

Fractal representation of nucleic acids could be useful for the detection of peculiarities in the sequence composition in short stretches embedded within longer sequences, and for guiding the manual alignment of nucleotide sequences. It is also conceivable that it could find applications in computerized multiple alignments as well as for the analysis of sequence homologies. For example, the degree of coincidence of barograms from related sequences could be used to derive dot matrices for sequence homology independent from nucleotide window sizes.

Also, one of the applications of graphic representation of nucleotide sequences is the use of image analysis for the screening of large sequences for regions of difference or homology. Most, if not all, computer programs presently available for this kind of analysis ultimately depend on the direct analysis of the complete nucleotide system. As the volume of sequenced genomes continues its exponential progression, new avenues should be explored for their analysis. If sequence input could be mapped to output images then computerized procedures for image analysis could be applied to genomic sequences.

Also, the kind of fractals obtained by the barogram transformation could be used in image synthesis in the same manner as other random fractals.[7,13,14] Thus, it could be technically feasible to create fractal landscapes based on the genetic sequence of a gene or a virus. Since many phenotypic characteristics of live beings can be considered as fractal renderings of their genetic information, given the proper transfer function (i.e., suitable models for genetic transcription, translation, and morphogenesis), the kind of images that could be derived from nucleotide sequences may be already familiar to most biologists.

References

1. Crilly, A.J., Earnshaw, R.A. and Jones, H., eds., *Fractals and Chaos* (Springer Verlag, 1991).
2. Hamori, E., "Novel DNA sequence representation", *Nature* **314**, 585 (1985) (Scientific correspondence).

3. Hurst, H.E., Black, R.P. and Simaika, Y.M., *Long Term Storage: An Experimental Study* (Constable, 1965).

4. Koutsky, L.A., Galloway, D.A. and Holnes, K.K., "Epidemiology of genital human papillomavirus infection", *Epidemiol. Rev.* **10**, 122–162 (1988).

5. Mandelbrot, B.B., *The Fractal Geometry of Nature* (W.H. Freeman and Co., 1982).

6. Minard, C.J., "Map portraying the losses suffered by Napoleon's army during the Russian campaign in 1812", cited by Tufte, E.R., in *The Visual Display of Quantitative Information* (Graphics Press, Cheshire, 1983).

7. Peitgen, H.-O. and Saupe, D., *The Science of Fractal Images* (Springer Verlag, 1988).

8. Pickover, C., "Spectrographic representation of globular protein breathing motions", *Science* **223**, 181 (1984a).

9. Pickover, C., "Frequency representation of DNA sequences: Application to a bladder cancer gene", *J. Molec. Graphics* **2**, 50 (1984b).

10. Pickover, C., "DNA vectorgrams: Representation of cancer gene sequences as movements along a 2-D cellular lattice", *IBM J. Res. Dev.* **31**, 111–119 (1987).

11. Pickover, C., *Computer, Pattern, Chaos, and Beauty* (St. Martin's Press, 1990).

12. Pickover, C., "DNA and protein tetragrams: Biological sequences as tetrahedral movements", *J. of Molecular Graph.* **10**, 2–6 (1992).

13. Saupe, D., "Algorithms for random fractals", *The Science of Fractal Images*, eds. M.F. Barnsley, R.L. Devaney, B.B. Mandelbrot, H.-O. Peitgen, D. Saupe and R.F. Voss (Springer-Verlag, 1988) pp. 71–136.

14. Saupe, D., "Random fractals in image synthesis", *Fractals and Chaos*, eds. A.J. Crilly, R.A. Earnshaw and H. Jones (Springer-Verlag, 1991) pp. 89–118.

15. Schroeder, M., *Fractals, Chaos, Power Laws. Minutes From an Infinite Paradise* (W.H. Freeman and Co., 1990).

16. Seedorf, K., Kraemmer, G., Duerst, M., Suhai, S. and Roewekamp, W.G., "Human papillomavirus type 16 DNA sequence", *Virology* **145**, 181–185 (1985).

17. Silverman, B.D. and Linsker, R., "A measure of DNA periodicity", *J. Theor. Biol.* **118**, 295–300 (1986).

Appendix

Since 1992 the visualization of nucleic acids as fractal structures has exploded, the literature has expanded rapidly and the fractal nature of nucleic acids has been the subject of major scientific controversies.

DNA walks and Chaos game representation (CGR) are the two major types of transforming functions used to cnovert the discrete nature of nucleotide sequences into functions amenable to analysis within the theory of stochastic processes with stationary increments. In general, CGR approaches[12,13] have been investigated for sequence visualization and pattern recognition[15,17,21,25,31] while 1- and 2-D DNA walks have been predominantly explored to assess long-range internal correlations.

One-dimensional DNA walks can be generated by an incremental variable that is associated to each nucleotide position i with the value 1 or -1 depending on whether position i is occupied by a purine or a pyrimidine. This approach was used to claim the existence of long-range power law relations in intron-containing genes and non-transcribed sequences but not in cDNA or intronless genes.[10,20,26,27,32] A generalized bidirectional Lévy-walk model leads to similar results.[5] In these studies the exponent of the root mean square fluctuation was used as a quantitative value related to the existence of these relations. The range of the long-range correlations was even extended over an entire yeast chromosome.[23] In some cases these correlations may be explained by the alteration of introns and exons.[24] Also, the validity of this kind of analysis was challenged on the basis that the assumption of stochastic stationarity is problematic and that a multipatch model for sequence composition also explains the findings.[14] The reasons for these discrepancies were analyzed for several natural sequences[18] and in one of them[6] computer simulations were used to analyze the nature of the long-range correlations; artificially generated sequences were generated with local base compositions similar to those found along the genome of the lambda-phage. The calculation of the fractal exponent for the artificially generated sequences was found to be similar to those of the natural genomic sequence from which it was concluded that the ordering of nucleotides in the natural lambda-phage genome follows trivial statistical characteristics that may cause apparent long-range correlations. However, as pointed by the same authors, this does not mean that all DNA sequences will have the same trivial stochastic characteristics.[6]

Two-dimensional pseudorandom walks have also been used to calculate fractal dimensions in nucleotide sequences.[2,3,16] Using the sandbox algorithm, the multifractal spectrum of several natural nucleotide sequences was shown to be significantly different from artifical sequences matched in length, single-base, and dimer-base frequency.[2,3] This was considered indicative of information due largely to a non-uniform distribution of bases and dimers within DNA sequences, i.e., information content not explained by single-base or dimer distribution. Regardless of the preferred interpretation (long-range correlations, patchiness, differences in local base composition, or induced based correlations), fractal dimensions represent new parameters reflecting nucleic acid composition and are potentially useful to measure and characterize groups of nucleotide sequences with respect to their taxonomic position and molecular evolution. Also, it has been indicated[33] that in DNA walks of the kind described above, the use of less than four dimensions introduces correlations between the bases. However, the existence of long-range correlations within nucleotide sequences of biological significance is also supported by findings using a large number of natural genetic sequences and binary algebra sequence transformations that do not introduce correlations between nucleotide bases.[22,23]

All phenomena, random or deterministic, exhibiting a fractional fractal dimension can be considered to be self-repetitive and fractal in nature.[30] As such a fractional Brownian motion generated by a perfectly random variable corresponds to a true random fractal process. Therefore, it is surprising that the finding of random fractal behavior in DNA Lévy walks[6,18] could have been proposed as evidence for the lack of fractalness in natural nucleotide sequences. Conversely, the fractal, i.e., internal repetitive, nature of DNA cannot be trivialized or dismissed on the basis that similar behavior can be generated by non-random evolutionary processes leading to internal repetitions.[14] The existence of deterministic, i.e., non-random, albeit chaotic processes underlying DNA fractalness could be explored with approaches similar to those used for detecting determinism in time series derived from a large variety of physical and biological processes.[1,7,8,9,11,19,28,29,30]

The fractal nature (internal repetitiveness) of nucleic acids appears to be firmly established. Similarly, there is no doubt that nucleotide biases exist in the composition of nucleotide sequences of biological significance.[4] However, controversy still exists on the nature of the deterministic processes underlying those biases as well as on whether fractal analysis of nucleotide sequences may help in the characterization of those processes.

References

1. Ben-Mizrachi, A., Procaccia, I. and Grassberger, P., "Characterization of experimental (noisy) strange attractors", *Physical Review A* **29** (1984).
2. Berthelsen, C.L., Glazier, J.A. and Skolnick, M.H., "Global fractal dimension of human DNA sequences treated as pseudorandom walks", *Physical Review A* **45**, 8902–8913 (1992).
3. Berthelsen, C.L., Glazier, J.A. and Raghavachari, S., "Effective multifractal spectrum of a random walk", *Physical Review E* **49**, 1860–1864 (1994).

4. Bronson, E.C. and Anderson, J.N., "Nucleotide composition as a driving force in the evolution of retroviruses", *J. Mol. Evol.* **38**, 506–532 (1994).

5. Buldyrev, S.V., Goldberger, A.L., Havlin, S., Peng, C.-K., Simons, M. and Stanley, H.E., "Generalized Lévy-walk model for DNA nucleotide sequences", *Physical Review E* **47**, 4514–4523 (1993).

6. Chatzidimitriou-Dreismann, C.A., Friedrich Streffer, R.M. and Larhammar, D., "Variations in base pair composition and associated long-range correlations in DNA sequences-computer simulation results", *Biochimica et Biophysica Acta* **1217**, 181–187 (1994).

7. Grassberger, P. and Procaccia, I., "Measuring the strangeness of strange attractors", *Physica* **9D**, 189–208 (1983).

8. Grassberger, P. and Procaccia, I., "Characterization of strange attractors", *Physical Review Letters* **50**, 346–349 (1984).

9. Grassberger, P., "Estimating the fractal dimensions and entropies of strange attractors", in *Chaos*, ed. A.V. Holden (Princeton University Press, 1986) pp. 291–311.

10. Grosberg, A.Yu., Rabin, Y., Havlin, S. and Nir, A., "Self-similarity in DNA structure", *Biofizika* **26**, 1–6 (1993).

11. Hentschel, H.G.E. and Procaccia, I., "Fractal nature of turbulence as manifested in turbulent diffusion", *Physical Review A* **27**, 1266–1269 (1983).

12. Jeffrey, H.J., "Chaos game representation of gene structure", *Nucleic Acids Research* **18**, 2163–2170 (1990).

13. Jeffrey, H.J., "Chaos game visualization of sequences", *Computers and Genetics* **16**, 25–33 (1992).

14. Karlin, S. and Brendel, V., "Patchiness and correlations in DNA sequences", *Nature* **259**, 677–680 (1993).

15. Korolev, S.V., Solovyev, V.V. and Tumanyan, V.G., "New method in the overall search for the functional regions of DNA using the fractal representation of nucleotide texts", *Biophysics* **37**, 733–742 (1992).

16. Korolev, S.V., Vin'a, A.R., Tumanyan, V.G. and Esipova, N.G., "Fractal properties of DNA sequences", in *Fractals in the Natural and Applied Sciences*, ed. M.M. Novak (Elsevier Science/North Holland, 1994).

17. Kumar Burma, P., Raj, A., Deb, J.K. and Brahmachari, S.K., "Genome analysis: A new approach for the visualization of sequence organization in genomes", *Journal of Biosciences* **17**, 395–411 (1992).

18. Larhammar, D. and Chatzidimitriou-Dreismann, C.A., "Biological origins of long-range correlations and compositional variations in DNA", *Nucleic Acids Research* **21**, 5167–5710 (1993).

19. Lauterborn, W. and Holzfuss, J., "Evidence for a low-dimensional strange attractor in acoustic turbulence", *Physical Letters A* **115**, 369–372 (1986).

20. Li, H.-H. and Kaneko, K., "Long-range correlations and partial $1/f \uparrow \alpha \uparrow$ spectrum in a non-coding DNA sequence", *Europhysical Letters* **17**, 655–658 (1992).

21. Lim, H.A. and Solovyev, V.V., "A new general approach for searching functional regions using fractal representation of nucleotide and amino acid sequences",

ACS Conference Proceedings Series, eds. M.R. Ladish and A. Bose (American Chemical Society, 1992) pp. 387–391.

22. Maddox, J., "Long-range correlations with DNA", *Nature* **358**, 103 (1992).

23. Munson, P.J., Taylor, R.C. and Michaels, G.S., "Long range DNA correlations extended over entire chromosome", *Nature* **360**, 636 (1992).

24. Nee, S., "Uncorrelated DNA walks", *Nature* **357**, 450 (1992).

25. Oliver, J.L., Bernaola-Galvan, P., Guerrero-Garcia, J. and Roman-Roldan, R., "Entropic prophiles of DNA sequences through chaos-game-derived images", *Journal of Theoretical Biology* **160**, 457–470 (1993).

26. Peng, C.-K., Buldyrev, S.V., Goldberger, A.L., Havlin, S., Sciortino, F., Simons, M. and Stanley, H.E., "Long-range correlations in nucleotide sequences", *Nature* **356**, 168–170 (1992).

27. Peng, C.-K., Buldyrev, S.V., Goldberger, A.L., Havlin, S., Sciortino, F., Simons, M. and Stanley, H.E., "Fractal landscape analysis of DNA walks", *Physics A* **191**, 25–29 (1992).

28. Roschke, J., "Eine Analyse der nichtlinearen EEG-Dynamik", Dissertation, Goettingen (1986).

29. Schaffer, W.M. and Kot, M., "Differential systems in ecology and epidemiology", *Chaos*, ed. A.V. Holden (Princeton University Press, 1986) pp. 158–178.

30. Schroeder, M., *Fractals, Chaos, Power Laws. Minutes from an Infinite Paradise* (W.H. Freeman and Co., 1990).

31. Solovyev, V.V., Lim, H.A., Milanesi, L. and Lawrence, C., "Application of fractal representation of genetic texts for recognition of genome functional and coding regions", *2nd International Conference on Bioinformatics, Supercomputing and Complex Genome Analysis*, eds. H.A. Lim, J.W. Fickett, C.R. Cantor and R.J. Robbins (World Scientific, 1992) pp. 609–622.

32. Stanley, H.E., Buldyrev, S.V., Goldberger, A.L., Hausdorff, J.M., Havlin, S., Mietus, J., Peng, C.-K., Sciortino, F. and Simons, M., "Fractal landscapes in biological systems: Long range correlations in DNA and interbeat heart intervals", *Physica A* **192**, 1–12 (1992).

33. Voss, R.F., "Evolution of long-range fractal correlations and $1/f$ noise in DNA base sequences", *Physical Review Letters* **68**(25), 3805–3808 (1992).

Glossary

fractal: A term coined by B. Mandelbrot to describe graphs, objects, functions, or patterns with self-similarity.

self-similarity: The property of fractals by which "zoomed-in" portions of a fractal still reproduce features similar to those observed at lower magnification. A trivial example could be the similarity of a branch of a tree with the tree itself. A magnified portion of a fractal would look qualitatively similar to the larger portion in which it is contained.

fractal dimension: Intuitively, this dimension is described as a measure of the degree with which a fractal fills the Euclidean dimensions in which it evolves. Formally, it is defined by the mathematical procedure permitting its calculation (see text).

radius of gyration: The mean quadratic distance from the center of mass in a given axis. It reflects an object's spatial extension and shape. It is also formally defined by the mathematical procedure leading to its calculation (see text).

papillomavirus: A group of viruses infecting several mammalian species. These viruses are non-enveloped and have an icosahedral capsid formed of structural proteins contained inside a double-stranded and circular DNA molecule of about 8000 nucleotides. Many of these viruses cause tumours in animals. Several types have also been related to genital cancers in humans. Type 16 is the one most commonly found in genital pre-cancerous and cancerous lesions.

Visualization of Open Reading Frames in mRNA Sequences

Perry B. Hackett, Mark W. Dalton, Darrin P. Johnson
and
Melvin R. Duvall

The DRAW algorithm is presented which permits the visualization of open reading frames (ORFs, RNA sequences that can be translated into protein) in messenger RNAs folded into low energy secondary structures. The ORFs are color-coded for easy localization, and the initiating AUG codons are color-coded to indicate the theoretical strength of their ability to initiate protein synthesis. The DRAW program allows display and plotting in either color or in different fonts and character sizes. The easy visualization of ORFs in folded mRNAs provides insight into potential sites of regulation of gene expression at the translational level.

Translation of mRNA

Studies of the translational efficiencies of many prokaryotic[1] and eukaryotic[2–8] mRNAs suggest that RNA structure, the formation of stems of base-paired RNA structures, and consequential loops may either enhance or reduce the rate of protein synthesis. RNA structure has particular importance in the expression of RNA tumor viruses (retroviruses) that cause cancer, leukemia, and AIDS. The full-length RNA of every animal retrovirus has four roles during the viral life cycle:

(1) replication — some of the RNA is a template which is copied into DNA;
(2) splicing — some of the RNA is spliced to form smaller mRNAs;
(3) translation — some of the RNA is directly translated to make required proteins for virus formation; and
(4) virus formation — some of the RNA is packaged into viruses for subsequent infection of other cells.

For each of these operations, either viral or cellular proteins and RNA interact with the full-length RNA to catalyze reactions. All of the functions described utilize regions in the 5'-leader RNA that precede the first gene sequence. Additionally, retroviral RNA structure appears to play several additional roles in human immunodeficiency virus (HIV) gene expression.[10]

The avian retroviruses like Rous sarcoma virus (RSV) have an additional complexity. The 5'-leader sequences of all mRNAs have three stretches of nucleotide sequences that can direct the synthesis of short polypeptides or proteins. These sequences are called open reading frames (ORFs) since they are *read* by ribosomes, the cellular machines that make proteins. ORFs begin with the nucleotide sequence AUG, a triplet of nucleotides that comprises the codon for methionine, and terminate with a termination codon (either UAG, UAA, or UGA) which does not specify an amino acid. In all of the avian retroviruses, the ORFs are conserved with respect to

(i) their positions in the leader RNA,
(ii) their lengths, and
(iii) the nucleotide contexts[11] of their AUG initiation codons, which determines their capability to initiate protein synthesis.

However, the encoded amino acid sequences in the ORFs are variable so that their conservation suggested a role in regulation of gene expression.[12] Several eukaryotic mRNAs, for genes involved with growth and development,[13] contain one or more short ORFs in their 5'-leader RNA sequences that have been implicated in regulation of gene expression.[14-16] Ribosomes translating one ORF may influence the RNA structure in another region of the mRNA.[17,18] Thus, RNA secondary structure may regulate translation by limiting access of coding regions to ribosomes and translation of ORFs in the 5'-leader RNA sequences may open up RNA structure in other regions to permit protein synthesis elsewhere.

To facilitate investigations of RNA structure on translation, we developed the DRAW algorithm which permits the visualization of ORFs in folded messenger RNAs. The ORFs are color-coded for easy localization and the initiating AUG codons are coded to indicate the theoretical strength of their initiation potential due to their flanking nucleotides.

Calculations of Secondary Structure

RNA secondary structure was predicted using an optimized version of the Zuker MFOLD algorithm that displays optimal and suboptimal folding patterns.[19] The program was run on a Silicon Graphics Iris 4D/240S using the Freier *et al.* thermodynamic parameters and the Jaeger loop parameters for predicting secondary structures.[20]

DRAW program

The DRAW program, written in Professional FORTRAN (version 1.00), utilizes the capabilities of the Graphics Development Toolkit (version 1.2) to produce output on an IBM 6180 eight-pen plotter. The program requires as input a file of two-dimensional spatial coordinates for each nucleotide representation. This file is generated by the MOLECULE program, originally written by R. Cedergren[21] and modified by J. Ryan Thompson (Carnegie-Mellon University). MOLECULE is available in the public Bionet directory (PC-SOFTWARE.THOMPSON). The DRAW program then generates a color-coded image of the mRNA sequence on the display, with additional user-controlled options for plotting in color or black-and-white.

Folding of mRNA sequences with short ORFs

Figure 1 shows the folding pattern of Rous sarcoma virus RNA, a model RNA, using the colorized version of DRAW. Panel A displays the folded structure of the first 600 nucleotide of the sequence of the PrC strain[22] of the Rous sarcoma virus and panel B shows the structure of a mutated RSV mRNA leader sequence, containing a stem of 81 base pairs, that was constructed as a control to measure the rate of internal initiation at the *gag* AUG initiation codon.[23] The 81-base paired stem has a ΔG of about -181 kcal/mole and is resistant to translation even though it contains two internal ORFs. An ORF with a

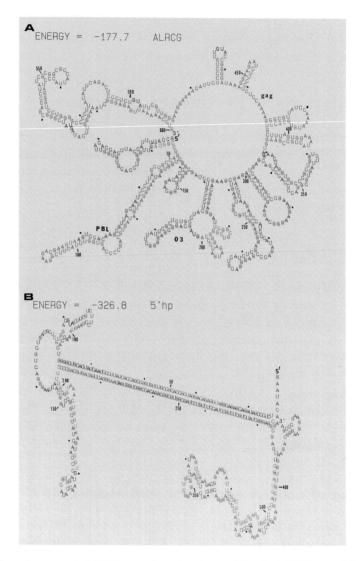

Fig. 1. Computer-generated RNA secondary structures of the first 600 bases of the PrC strain of Rous sarcoma virus RNA (GenBank designation, ALRCG) and the first 424 bases of 5′-hp RNA plotted with the colorized DRAW program. Color coding for the ORFs in the leader RNA sequences are described in the text. Structural motifs include: PBL, tRNA[Trp] Primer Binding Loop; O3, ORF-3 Loop; *gag*, the Pr76*gag* gene of RSV which begins at nucleotide 380. The 5′ end of each sequence is marked with a large 5′. Dots are placed every 20 nucleotide and sites are marked every 50 bases. (A) RSV RNA and (B) 5′-hp RNA.

moderate AUG codon (bases 31-33) is buried in the stem; 19 base pairs precede the AUG codon, making penetration by the scanning ribosome difficult. Translation downstream from the major stem is reduced more than 90%.[23] In both folds the untranslated sequences are in orange and sequences potentially subject to translation are in blue. AUG codons are color-coded, green represents an AUG of moderate translational initiation potential [U--**AUGG**, representing any base], purple represents an AUG codon of somewhat poorer initiation strength [A--**AUGA**], red indicates the theoretically strongest initiation signal [A--**AUGG**], and brown represents either an extremely poor initiator AUG signal [Py--**AUG**Py, Py = U or C] or an internal AUG codon. Although the AUG codons are ranked according to their flanking nucleotides, other features influence initiation strength.[24] Initiation of retroviral protein synthesis is at the red AUG codon (nucleotides 480–482) preceding the long ORF that continues for hundreds of bases beyond the sequence shown.

In contrast, the same structures in Fig. 1 are presented using the the black-and-white version of DRAW in Fig. 2. Here, the untranslated regions are in small capital letters, the short ORFs behind less than ideal AUG codons are designated by large lower case letters, and the *gag* gene ORF behind a strong AUG initiation codon is displayed in large capital letters.

Inspection of either of the RSV RNA structures in Figs. 1(a) and 2(a) reveals a major structure, the PBL (Primer Binding Loop) stem/loop, in the leader RNA that lies between the 5' end of the RNA and this protein initiation site for the *gag* gene product. The calculated energy, ΔG, of the PBS stem/loop is -60.3 kcal/mole which is sufficiently stable to prevent ribosomal subunits from efficiently scanning the leader sequence. However, owing to the strategic placement of the first two ORFs in the leader RNA, scanning ribosomal subunits have the opportunity of converting to full ribosomes at the ORFs and thereby penetrate and break open the PBL stem/loop. For both the RSV structure and the 5'-hp structure simple visualization of the ORFs greatly assists in evaluation of the primary and secondary structures of the mRNAs and their potential roles in translational regulation.

The folded structure of the mRNA encoding the receptor protein for thyroid hormone (Fig. 3) shows a third example of the strategic placement of an ORF. The gene for one of the thyroid hormone receptors is termed c-erbAβ [25] owing to its homology to a viral gene that causes erythroblastosis. In this case the ORF, in a highly stable stem structure, encodes a peptide of 43 amino acids. The ORF in the c-erbAβ mRNA stretches between nucleotides 11 and

ENERGY = -153.8   RATCERBAB

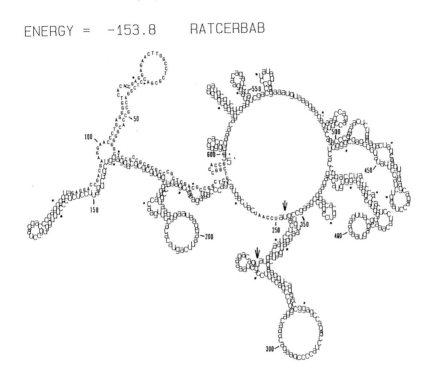

Fig. 3. RNA secondary structure of the first 600 bases of the thyroid hormone receptor gene transcript using the black-and-white DRAW program. The conventions for representing ORFs in the RNA are the same as those shown in Fig. 2. Two possible initiation codons for thyroid hormone receptor synthesis are indicated by the arrows at positions 251-253 and 266-268.

236 in the leader sequence. This mRNA is difficult to translate in a cell-free system. The stability of the leader structure, $\Delta G < -80$ kcal/mole, suggests that the mRNA would be infrequently translated. Yet the thyroid hormone receptor protein is made in significant quantities. The leader ORF with an AUG codon (nucleotides 116–118) of moderately strong initiation strength suggests a mechanism by which the mRNA can be utilized for protein synthesis. As with the RSV mRNAs, the ORF in the c-erbAβ is situated such that its translation would break apart most of the secondary structure in the leader to permit ribosomes to traverse the leader and reinitiate at either the AUG codon at nucleotides 251–253 or at a second site at 266–268. The second AUG codon appears to be the major site of translational initiation.[25]

Summary

Two versions of the DRAW algorithm are presented. The colorized version is the most practical for easy visualization of ORF location and AUG codon strength. However, for publication, where the cost of color plates is high, the black-and-white version is economical but limited by the range of available fonts and sizes that can be easily distinguished. Both should be useful for analyses of mRNA structure/function relationships in translation and other reactions involving mRNA.

Acknowledgments

We thank Michael Zuker for his help and encouragement in our efforts to understand the importance of RNA structure in translation and Bob Petersen, Jeff Essner, and Howard Towle for discussions and/or reviewing the manuscript. This work was supported by NIH grants CA 43881 and RR 06625.

References

1. deSmit, M.H. and van Duin, J., "Control of prokaryotic translational initiation by mRNA secondary structure", *Prog. Nuc. Acid Res. Mol. Biol.* **38**, 1–35 (1990).

2. Kozak, M., "The scanning model for translation: An update", *J. Cell. Biol.* **108**, 229–241 (1989).

3. Pavlakis, G.N. Lockard, R.E., Vamvakopoulos, N., Rieser, L., RajBhandary, U.L. and Vournakis, J.N., "Secondary structure of mouse and rabbit α- and β-globin mRNAs: Differential accessibility of α and β initiator AUG codons towards nucleases", *Cell* **19**, 91–102 (1980).

4. Lowe, W.L., Roberts, C.T., Lasky, S.R. and LeRoith, D., "Differential expression of alternative 5' untranslated regions in mRNAs encoding rat insulin-like growth factor I", *Proc. Natl. Acad. Sci. U.S.A.* **84**, 8946–8950 (1987).

5. Godefroy-Colburn, T., Ravelonandro, M. and Pinck, L., "Cap accessibility correlates with initiation efficiency of alfalfa mosaic virus RNAs", *Eur. J. Biochem.* **147**, 549–552 (1985).

6. Jobling, S.A. and Gehrke, L., "Enhanced translation of chimeric messenger RNAs containing a plant untranslated lead sequence", *Nature* **325**, 322–325 (1987).

7. Dabrowski, C. and Alwine, J.C., "Translational control of synthesis of simian virus 40 late protein from polycistronic 19S late mRNA", *J. Virol.* **62**, 3182–3192 (1988).

8. Klausner, R.D. and Harford, J.B., "*cis-trans* models for post-transcriptional gene regulation", *Science* **246**, 870–872 (1989).

9. Jacks, T., Power, M.D., Masiarz, F.R., Luciw, P.A., Barr, P.J. and Varmus, H.E., "Characterization of ribosomal frameshifting in HIV-1 *gag-pol* expression", *Nature* **331**, 280–283 (1988).

10. Pavlakis, G.N. and Felber, B.K., "Regulation of expression of human immunodeficiency virus", *New Biol.* **2**, 20–31 (1990).

11. Kozak, M., "An analysis of 5'-noncoding sequences from 699 vertebrate messenger RNAs", *Nuc. Acids Res.* **15**, 8125–8148 (1987).

12. Hackett, P.B., Petersen, R.B., Hensel, C.H., Albericio, F., Gunderson, S.I., Palmenberg, A.C. and Barany, G., "Synthesis *in vitro* of a seven amino acid peptide encoded in the leader RNA of Rous sarcoma virus", *J. Mol. Biol.* **190**, 45–57 (1986).

13. Kozak, M., "Bifunctional messenger RNAs in eukaryotes", *Cell* **47**, 481–483 (1986).

14. Petersen, R.B., Moustakas, A. and Hackett, P.B., "A mutation in the short 5'-proximal open reading frame of Rous sarcoma virus RNA alters virus production", *J. Virol.* **63**, 4787–4796 (1989).

14a. Moustakas, A., Sonstegard, T. and Hackett, P.B., "Alterations of the three upstream open reading frames inthe Rous sarcoma virus leader RNA modulate viral replication and gene expression", *J. Virol.* **67**, 4337–4349 (1993).

14b. Moustakas, A. and Hackett, P.B., "Effects of the open reading frames in the Rous sarcoma virus leader RNA on translation", *J. Virol.* **67**, 4350–4357 (1993).

15. Mueller, P. and Hinnebushch, A.G., "Multiple upstream AUG codons mediate translational control of GCN4", *Cell* **45**, 201–207 (1986).

16. Petersen, R.B.and Hackett, P.B., "Characterization of ribosome binding on Rous sarcoma virus RNA *in vitro*", *J. Virol.* **56**, 683–690 (1985).

17. Fajardo, J.E. and Shatkin, A.J., "Translation of bicistronic viral mRNA in transfected cells: Regulation at the level of elongation", *Proc. Natl. Acad. Sci. USA.* **87**, 328–332 (1990).

18. Shatkin, S.H. and Liehaber, S.A., "Destabilization of messenger RNA/complimentary DNA duplexes by the elongating 80S ribosome", *J. Biol. Chem.* **261**, 16018–16025 (1989).

19. Zuker, M., "On finding all suboptimal foldings of an RNA molecule", *Science* **244**, 48–52 (1989).

20. Jaeger, J.A., Turner, D.H. and Zuker, M., "Improved predictions of secondary structures for RNA", *Proc. Natl. Acad. Sci. USA* **86**, 7706–7710 (1989).

21. Lapalme, G., Cedergren, R.J. and Sankoff, D., "An algorithm for the display of RNA secondary structure", *Nuc. Acids Res.* **10**, 8351–8356 (1982).

22. Schwartz, D.E., Tizard, R. and Gilbert, W., "Nucleotide sequence of Rous sarcoma virus", *Cell* **32**, 853–869 (1983).

23. Hensel, C.H., Petersen, R.B. and Hackett, P.B., "Effects of alterations in the leader sequence of Rous sarcoma virus RNA on initiation of translation", *J. Virol.* **63**, 4986–4990 (1989).

24. Hackett, P.B., Dalton, M.W., Johnson, D.P. and Petersen, R.B., "Phylogenetic and physical analysis of the 5'-leader RNA sequences of avian retroviruses", *Nuc. Acids Res.* **19**, 6929–6934 (1992).

25. Murray, M.B., Zilz, N.D., McCreary, N.L., MacDondald, M.J. and Towle, H.C., "Isolation and characterization of rat cDNA clones for two distinct thyroid hormone receptors", *J. Biol. Chem.* **263**, 12770–12777 (1988).

Glossary

eukaryote (eukaryotic): A cell that has a nucleus, e.g., an animal or plant cell.

RNA leader sequence: The sequence of nucleotides between the beginning of a messenger RNA molecule and the primary sequence that encodes protein.

prokaryote (prokaryotic): A cell that does not have a nucleus, e.g., a bacterium.

retrovirus (retroviral): A virus that has an RNA chromosome which is copied into a complementary DNA sequence following infection of a host cell. Retroviruses are also known as RNA tumor viruses.

ORF: A ORF (open reading frame) is a stretch of nucleotide sequences which has the potential to code for a protein, but for which there is no proof of actual production of the said protein. The stretch of bases do not contain any termination codons. Codons are triplets of bases that code for amino acids. Termination, or stop, codons (TAA, TAG, or TGA) do not code for amino acids but rather stop the addition of amino acids to a growing polypeptide. Since the genetic code is read in sets of three bases without punctuation, it can be read in one of the three ways. Two of the ways are incorrect and it may happen that the particular set of bases in one of these two frames codes for a termination signal (e.g., TGA). This would stop protein synthesis and "close" the reading frame.

Visualization of Protein Sequences Using the Two-Dimensional Hydrophobic Cluster Analysis Method

Michel T. Semertzidis, Etienne Thoreau
Anne Tasso, Bernard Henrissat, Isabelle Callebaut
and
Jean Paul Mornon

This paper describes a method for displaying protein sequences in the form of 2-D graphics, known as the "HCA method" (HCA). This technique, essentially visual, makes it possible to compare and align reliably protein sequences that exhibit less than 20% similarities. HCA may also be of interest in predicting secondary structures from amino acid sequences. The usefulness of the method is demonstrated through a series of examples extracted from the most recent literature. A discussion of the superiority of helical nets over other 2-D plots is included.

Introduction

With the rapid expansion of the biological data banks, the design of efficient methods to extract the structural and functional information contained in protein sequences becomes of utmost importance. This has led to the emergence of a number of methods to compare and analyze the primary structure of proteins. In this context, we have proposed a new method, the Hydrophobic Cluster

Analysis (HCA), based on a helical representation of protein sequences.[5,8] HCA
is not primarily based on the maximization of a homology score, but rather on
the visual detection and comparison of clusters of hydrophobic residues formed
in the 2-D helical net underlying the method. In this chapter, we illustrate
with new examples how the display of sequences in helical nets can be useful by combining (i) the efficient eye-brain perception of shapes and (ii) the
structural principles that govern protein architecture. Here, we also introduce
statistical analysis of the hydrophobic cluster characteristics.

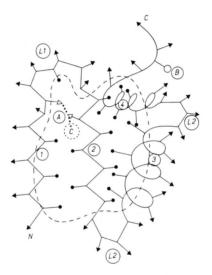

Fig. 1. Schematic representation of a globular protein domain. Frequently observed structures are: (1) edge β strands (made of •'s, which are essentially hydrophobic residues and
▲'s, which are essentially hydrophilic residues), (2) internal β strands; (3) edge amphiphilic
α helices, i.e., helices possessing a hydrophilic face directed towards the solvent and an opposite hydrophobic face inserted in the hydrophobic core; (4) internal α helices (rare), (L1)
loops containing hydrophobic residue(s), and (L2) mostly hydrophilic loops sometimes including β turn(s). The hydrophobic core is delineated by the dotted line. Rare situations
may occur as in (a), where two hydrophilic residues neutralize each other through hydrogen bond(s) or a salt bridge, or as in (b) where an hydrophobic residue is isolated in an
hydrophilic environment. Water surrounds the globular domain, and can occupy internal
niche(s) as in (c). The ratio of hydrophobic/hydrophilic residues used in this scheme (about
1/3) is similar to that observed for typical globular domains. When two or more strands are
linked by regular main-chain N-H...O=C H-bonds (parallel or antiparallel), they form a β
sheet as possibly antiparallel strands 1 and 2 in this figure.

HCA Principles

The reader can find a detailed description of the principles underlying the HCA method in a review by Lemesle-Varloot *et al.* (1990). Therefore, in the current chapter, we only summarize the fundamental points.

All globular domains of proteins are characterized by a clear partition of two main groups of amino acids (Fig. 1): The hydrophobic ones which tend to cluster in the core of the protein to escape solvent, and the hydrophilic ones which constitute the surface of the protein that is in contact with water (all real globular proteins on Earth are completely dependent on the water environment). It is now generally considered that one of the main early driving forces for protein folding is the creation of a hydrophobic core (about one-third of the amino acids) harvesting almost all strong hydrophobic amino acids side chains (V, L, I, F), medium hydrophobic amino acids side chains (W, M, Y), and part of mimetic ones (A, C, T, S), i.e., amino acids which could as well occupy hydrophilic or hydrophobic sites. Consequently the distribution of hydrophobic amino acids along sequences is of crucial importance for characterizing the corresponding protein fold.

The idea of displaying sequences through an α helical net (HCA plots — Fig. 2), proposed several times in the sixties and seventies[3,11] and curiously each time forgotten, appears to be a very efficient way to visualize information relevant to the secondary and tertiary structural organization of proteins and particularly that of globular domains. Indeed, we are now able to demonstrate on a wide scale (all the Protein Data Bank entries, in preparation) the following points:

— Strong and medium hydrophobic amino acids are grouped in HCA plots on well-defined clusters.
— Centers of regular secondary structures (β strands and α helices), which constitute the building blocks of protein architecture, statistically coincide with the centers of hydrophobic clusters. Between hydrophobic clusters, one finds stretches of sequences concerning loops generally characterized by hydrophilic amino acids (D, E; N, Q; H, K, R), mimetic amino acids (A, C, T, S), and special ones (G and P which bring either the greatest flexibility or the largest constraint to the main chain, respectively) (cf., statistical analysis of reference 9).

The α helical net used in the HCA method is the most useful net for presentation sequence information, although closely related nets can also be used.

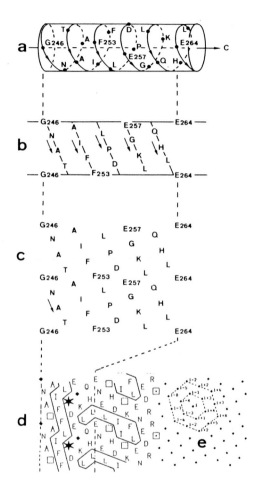

Fig. 2. 2-D α helical patterns for HCA analysis. The example is the G246...E264...S283 part of the human α1-antitrypsin sequence GNATAIFFLPDEGKLQHLE.... The sequences are written on a classical α helix (3.6 amino acids per turn) smoothed on a cylinder. After five turns, residues i and $i + 18$ (e.g., G246 and E264) have similar positions parallel to the axis of the cylinder (2a). To make the 3-D representation easy to interpret and work with, the cylinder is then cut parallel to its axis (e.g., along G246 and E264) and unrolled (2b). As some adjacent amino acids (e.g., F253 and L254) are now separated by the unfolding of the cylinder, the representation is duplicated to restore the full connectivity of each amino acid (2c). Sets of adjacent hydrophobic residues (VLIFMYW) are then contoured and named hydrophobic clusters (2d). Special symbols are used for P (★), G (♦), S (⊡) and T (□) (see Fig. 3). Figure 2(e) shows the environment of first and second neighbors around residue i, respectively $i - 4$ to $i + 4$ and $i - 8$ to $i + 8$.

At any level of sequence identity, similar HCA plots for globular domains indicate similar 3-D folds. The HCA is a very efficient way to extract structural and functional information from sequences (e.g., isolate sequences, several sequences or, better, from families of related sequences (cf., working strategies in reference 8).

Examples

The remarkable ability of helical nets to transform linear information into shapes and patterns easily perceptible by the human eye is shown in Fig. 3 where the amino acid sequence of a prion, a protein with unknown 3-D structure, is shown both in the conventional linear 1-letter code (Fig. 3(a)) and in the 2-D helical plot used in HCA (Fig. 3(b)). The HCA plot obviously carries a lot of information that can be visually analyzed. For example, a segmentation in five areas is rapidly performed as shown in Fig. 3(b). Segment 1, mainly hydrophobic, probably corresponds to the signal peptide. Domain 1 appears curiously constituted by the repetition or the quasi-repetition of several amino acids motifs (W, Y, P, G, Q). It is obviously not folded as a classical globular domain, i.e., with the characteristic regular succession of hydrophobic clusters (regular secondary structures) separated by loops. Such a regular segmentation is found in domain 2, downstream segment 2 which is likely a hinge region separating domains 1 and 2 (note the accumulation of alanine and glycine (one main constituant of loops)). At the C-terminus occurs a characteristic transmembrane stretch of twenty amino acids, either fully hydrophobic or mimetic (S, T, P, G). Within domain 2, one can note an unusual frequency of tyrosines, a weakly hydrophobic residue which is frequently encountered in loops[9] as well as in the hydrophobic core. This should be taken into account when performing the structural segmentation of domain 2 into regular secondary structures and loops and/or when predicting these secondary structures (α helices or β strands). For all these points, comparison of several related sequences (if available) is always very helpful.

Figure 3(c) illustrates part of the information contained in the human Prion gene.[10] Plots (a), (b), (c), and (d) are relative to the translation of DNA strands as schematized in the accompanying box. Plot (b), displayed as a reference, is the normal amino acid sequence of human Prion (very similar to that of mouse Prion illustrated in Fig. 3(b)). Vertical dotted lines indicate the same segmentation as for mouse Prion.

A MANLGYWLLALFVTMWTDVGLCKKRPKPGGWNTGGSRYPGQGSPGGNRYP
PQGGTWGQPHGGGWGQPHGGSWGQPHGGGWGQPHGGGWGQGGGTHNQWNK
PSKPKTNFKHVAGAAAAGAVVGGLGGYMLGSAMSRPMIHFGNDWEDRYYR
ENMYRYPNQVYYRPVDQYSNQNNFVHDCVNITIKQHTVVTTTKGENFTET
DVKMMERVVEQMCVTQYQKESQAYYDGRRSSSTVLFSSPPVILLISFLIF
LIVG

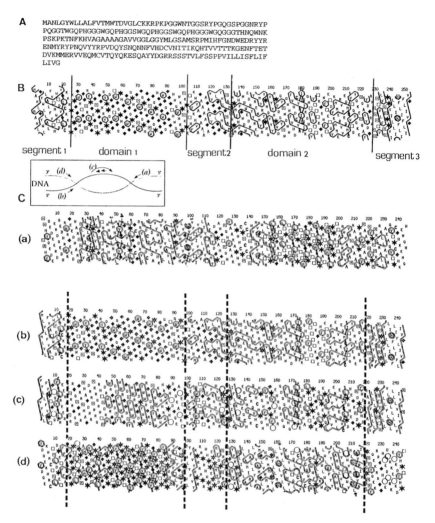

Fig. 3. HCA plots of the Prion, a protein with an unknown 3-D structure. (a) Amino acid sequence (one letter code) of the mouse prion protein.[19] (b) Same sequence displayed as an HCA plot (cf., Fig. 2 for principles), four amino acids are conventionally represented by special symbols (★: proline; ◆: glycine; □: threonine; ▣: serine). When possible, a further visual enhancement is achieved by the use of the following colours in the plots: Green for hydrophobic residues (VLIFWMY), red for "acidic" (DENQ) and proline (P) residues, blue for "basic" residues (HKR), and black for the other residues (GSTCA). (c) Gene information concerning human Prion; within the box are indicated the reading sense (a), (b), (c), and (d) of DNA strands (see text). Generated stop condons are indicated by open circles.

Plot (c) is that of the artificial sequence obtained by reading each codon in the reverse sense. Generated stop codons are indicated by open circles. Such plots illustrate the general codons property to code for same kinds of amino acids when read in both senses ("symmetry" around the second base). Consequently, HCA plots of normal globular domains characterized by the regular succession of hydrophobic clusters and loops are conserved in such an operation. Here, this obviously occurs for domain 2 and not as expected for domain 1. One can find another illustration of this property in Fig. 11 in reference 8.

Plot (d) illustrates another property of the code: The complementary codons of glycine are the codons for proline and vice versa; these two amino acids are special since they give more and less freedom to the main chain, respectively. Finally, plot (a) shows the amino acid sequence of an unexpected open reading frame (ORF) discovered in the entire length of the human Prion complementary strand, possibly leading to the expression of an "anti-Prion" protein.[6] One can compare the difference in the information contained in the HCA plots (a) and (b), relatively to that shown for the same sequences through 1-D profiles.[6]

In particular cases, such as that of the laminin B2 illustrated in Fig. 4, it can be useful to integrate mimetic residues (A, C, S, T) into clusters in order to better visualize the associated secondary structure elements. Incorporation

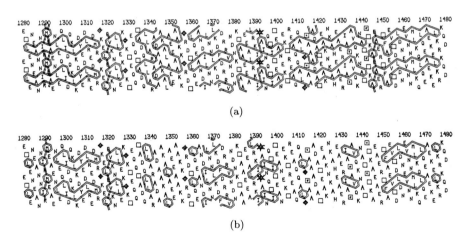

Fig. 4. HCA plot of a part of the laminin B2 sequence[12] with incorporation (a) or no incorporation (b) of the alanine residues within hydrophobic clusters.

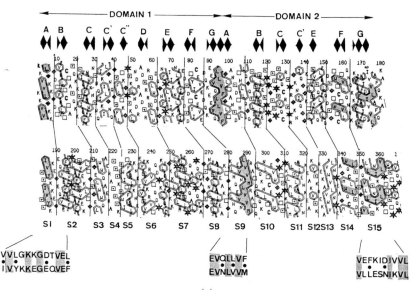

(a)

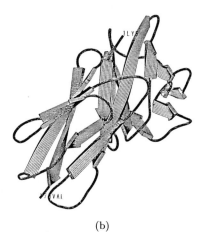

(b)

Fig. 5. (a) HCA plot of the extracellular part of human CD4 split to show the internal symmetry. Vertical lines delineate the structural segmentation (S1 to S15) within the two immunoglobulin-like domains. The secondary structure of domains 1 and 2 is indicated above their HCA plot according to the description of the 3-D structure.[15,18] β strands are depicted as $\blacklozenge$. Shaded hydrophobic clusters indicate anchor points for the sequences alignment. Under the HCA plots is some sequence alignment information characterizing the homology. (b) Cartoon 3-D representation of the two first immunoglobulin-like domains of CD4.[13]

of the laminin B2 alanine residues into clusters allows a clearer visualization of its helical content (horizontally shaped clusters).

The CD4 example (Fig. 5(a)) further illustrates the power of the HCA method which is able to detect structural homologies with level of sequence identities < 10%. In this particular case, HCA supports the hypothesis that CD4 may have evolved by gene duplication of a segment composed of two immunoglobulin-like domains[20] and whose structure has been solved (cf., Fig 5(b)).[15,18] Such repetition of divergent Ig-like domains has already been detected with HCA for cytokine receptors and thereafter verified.[17]

Figure 5(a) gives evidence for the internal symmetry within the extracellular part of CD4: Indeed, segmentation of the two 200 residues segments into loops and hydrophobic clusters reveals a similar distribution in the two segments. Moreover, the shape similarity of corresponding clusters, which is evident, for example, within segment 1, 9, 14, 15, seems indicative of similar secondary and tertiary structures. Values for HCA scores (68%) and sequence identities (16%) are within the range of significance observed for structurally related domains.[8]

The 3-D structure of the 70K HSC protein[4] shows a surprising similarity at the 3-D level with those of hexokinase and actin[7] although the sequence identity is very low. This example has been proposed as reference by Bowie *et al.* (1991) to test a powerful methodology which operates through several similar principles as HCA but in a reverse way. In Fig. 6, HCA of actin and HSC70

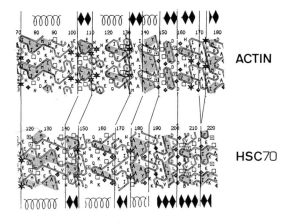

ACTIN

HSC70

Fig. 6. Part of the HCA plots of actin and HSC70 giving evidence of the structural homology existing between the two proteins.

indicate possibly related domains, as first recognized by the comparison of their 3-D structures.

Statistical Analysis of the Hydrophobic Clusters Characteristics

The purpose of these studies was to evaluate the effectiveness of the representation of protein sequences through helical nets rather than through any other geometric structure. To this end we have done comparative studies on groups of protein sequences of known structure, focusing on the characteristics of the hydrophobic clusters generated by different helicoidal representations.

Research for the best two-dimensional support for visualization of protein sequences and hydrophobic clusters

Linear- or β-sheet-like representation

In this representation, two problems arise for which there is no obvious solution. The length(s) of segments composing strands is (are) arbitrary as well as the choice of the residue located at the beginning of the first strand. Therefore, we cannot determine the neighbor rules between amino acids in this particular 2-D representation of sequences.

Helical representation

Helices overcome these problems because they are uniformly periodical. We have considered several types of helices, three of them being summarized in Table 1, searching for the "best" net, i.e., that carrying the best information contrast between real and random sequences. The environment of every amino acid is defined according to the principles of the HCA method: For example, in the α helix, the proximate environment of the ith residue is composed of the residues $i-4$, $i-3$, $i-1$, $i+1$, $i+3$, $i+4$, with $i-2$ and $i+2$ in addition if we consider mosaic clusters (see, for example, Fig. 3 around position 180). These mosaic clusters are often associated with β strand structures.

Table 1.

Helix type	Natural occurrence	Residues per turn	Rise per residue along the axis	Radius of helix (A°)	NEIGHBOR RULES first neighbors of residue i (first and second neighbors of residue i)
3–10	small pieces	3.0	2.0	1.9	$i \pm (1, 2, 3)$ $(i \pm (1, 2, 3, 4, 5, 6))$
α	abundant	3.6	1.5	2.3	$i \pm (1, 2, 3, 4)$ $(i \pm (1, 2, 3, 4, 5, 6, 7, 8))$
β	unobserved	4.4	1.1	2.8	$i \pm (1, 3, 4, 5)$ $(i \pm (1, 2, 3, 4, 5, 6, 7, 8, 9, 10))$

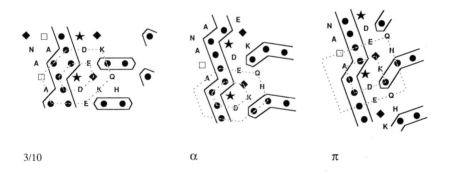

3/10 α π

Fig. 7. Schematic representation of three HCA plots using different types of helices (3/10, α, π).

The form of clusters varies as a function of the two-dimensional support type (see Fig. 7).

Do the clusters carry information relevant to the three-dimensional structure of proteins?

If the hydrophobic clusters are somehow reflecting the folding characteristics of proteins, a significant difference should be found between the clusters observed in natural proteins and those created by random sequences (most of them probably leading to unstable folds in water).

Random sequences were generated by random permutations of the amino acids in the real sequences. We have compared the distributions of the cluster lengths observed for real proteins with those found in the random sequences. The analysis was carried out on three classes of proteins with known 3-D structures: all α (67 proteins, 11235 residues), all β (76 proteins, 14914 residues), and α/β (22 proteins, 5418 residues). Six types of helical representations have been evaluated including those listed in Table 1. Significance of the results was first evaluated by the $\chi 2$ test.[16]

It appears that the α helical plot shows the best contrast between real and random sequences; consequently, it probably carries the largest amount of information concerning sequence/3-D structure relationships.

The reader can find confirmation of these facts in references 21 and 22.

Conclusion

During the last few years, many HCA applications have been successfully performed and several new ones are in press or submitted. On average, we have observed that about two-thirds of the studies on proteins showing sequence identities in the 10% to 20% range have led to the determination of accurate structural and/or functional information by HCA (e.g., reference 14). In the frequent cases where HCA has found a similarity with protein(s) having a known 3-D structure, complete 3-D models could be designed by homology modeling. So, although apparently time consuming, HCA analysis of protein sequences is always rewarding, and should have important consequences for future analyses of sequence information.

HCA software

Hydrophobic Cluster Analysis of protein sequences is performed using several programs producing HCA representations on paper and/or on-screen, for micro- or mini-computers. These programs are now commercially available from Doriane Company, 16 rue Julien Poupinet, F78150 Le Chesnay, France, Fax: (33-1) 39 55 57 82. Automation of HCA is underway and will generate new versions of HCA software in the future.

Note

The cytokine receptor family typically illustrates the power of the HCA method: The structural prediction,[17] has been completely verified by the determination of the 3-D structure of the human growth hormone/growth hormone receptor complex.[2] A further study of the Perion molecule has been published by the authors[2,3] and independently verified.[24]

Acknowledgments

This work was supported by CNRS, INSERM (contracts 889009 and 910912), Fondation pour la Reherche Médicale, Organibio (Organisation Nationale Interprofessionnelle des Bioindustries). Semertzidis and Thoreau have a Ph.D. grant from the French government (MRT) and from Institut Scientifique Roussel, respectively. Tasso is a post-doctoral fellow of Organibio and Callebaut is Research Assistant at the Fonds National de la Recherche Scientifique (FNRS-Belgique).

References

1. Bowie J.U., Lüthy, R. and Eisenberg, D., "A method to identify protein sequences that fold into a known three-dimensional structure", *Science* **253**, 164–170 (1991).
2. De Vos, A.M., Ultsch, M. and Kossiaakoff, A.A., "Human growth hormone and extracellular domain of its receptor: Crystal structure of the complex", *Science* **255**, 306–312 (1992).
3. Dunhill, P., "The use of helical net diagrams to represent protein structures", *Biophys. J.* **8**, 865–875 (1968).
4. Flaherty, K.M., DeLuca-Flaherty, C. and Mc Kay, D.B., "Three-dimensional structure of the ATPase fragment of a 70K heat shock cognate protein", *Nature* **346**, 623–628 (1990).
5. Gaboriaud, C., Bissery, V., Benchetrit, T. and Mornon, J.P., "Hydrophobic cluster analysis: An efficient new way to compare and analyse amino acid sequences", *FEBS Lett.* **224**, 149–155 (1987).
6. Goldgaber, D., "Anticipating the anti-prion protein", *Nature* **351**, 106 (1991).
7. Kabsch, W., Mannherz, H.G., Suck, D., Pai, E.F. and Holmes, K.C., "Atomic structure of the actin: DNaseI complex", *Nature* **347**, 37–44 (1990).
8. Lemesle Varloot, L., Henrissat, B., Gaboriaud, C., Bissery, V., Morgat, A. and Mornon, J.P., "Hydrophobic cluster analysis: Procedures to derive structural and functional information from 2-D representation of protein sequences", *Biochimie* **72**, 555–574 (1990).
9. Leszczynski, J.F and Rose, G.D., "Loops in globular protein: A novel category of secondary structure", *Science* **234**, 849–855 (1986).
10. Liao, Y.C.J., Lebo, R.V., Clawson, G.A. and Smuckler, E.A., "Human prion protein cDNA: Molecular cloning, chromosomal mapping and biological implica-

tions", *Science* **233**, 364–367 (1986).

11. Lim, V.I., "Polypeptide chain folding through a highly helical intermediate as a general principle of globular protein structure formation", *FEBS Lett.* **89**, 10–14 (1978).

12. Pikkareinen, T., Kallunki, T. and Trygguason, K., "Human laminin B2 chain. Comparison of the complete amino acid sequence with the B1 chain reveals variability in sequence homology between different structural domains", *J. Biol. Chem.* **263**, 6751–6758 (1988).

13. Priestle, J.P., "A stereo cartoon drawing program for proteins", *J. Appl. Cryst.* **21**, 572–576 (1988).

14. Py, B., Bortoli-German, I., Haiech, J., Chippaux, M. and Barras, F., "Cellulase EGZ of *Erwinia chrysanthemi*: Structural organization and importance of His 98 and Glu 133 residues for catalysis", *Prot. Eng.* **4**, 325–333 (1991).

15. Ryu, S.E., Kwong, P.D., Truneh, A., Porter, T.G., Arthos, J., Rosenberg, M., Dai, X., Xuong, N., Axel, R., Sweet, R.W. and Hendrickson, W.A., "Crystal structure of an HIV-binding recombinant fragment of human CD4", *Nature* **348**, 419–426 (1990).

16. Semertzidis, M.T., "Contribution au développement de la méthode HCA: Le tracé alpha-hélicoïdal contient-il une information sur le repliement protéique?" Analyse statistique, DEA, Biomathématiques, Université de Paris VII (1991).

17. Thoreau, E., Petridou, B., Kelly, P.A., Djiane, J. and Mornon, J.P., "Structural symmetry of the extracellular domain of the cytokine/growth hormone/prolactin receptor family and interferon receptors revealed by hydrophobic cluster analysis", *FEBS Lett.* **282**, 26–31 (1991).

18. Wang, J., Yan, Y., Garrett, T.P.J., Liu, J., Rodgers, D.W., Garlick, R.L., Tarr, G.E., Husain, Y., Reinherz, E.L. and Harrison, S.C., "Atomic structure of a fragment of human CD4 containing two immunoglobulin-like domains", *Nature* **348**, 411–418 (1990).

19. Westaway, D., Goodman, P.A., Mirenda, C.A., Mc Kinley, M.P., Carlson, G.A. and Prusiner, S.B., "Distinct prion proteins in short and long scrapie incubation period mice", *Cell* **51**, 651–662 (1987).

20. Williams, A.F., Davis, S.J., He, Q. and Barclay, A.N., "Structural diversity in domains of the immunoglobulin superfamily", *Cold Spring Harb. Symp. Quant. Biol.* **54**, 637–647 (1989).

21. Woodcock, S., Mornon, J.P. and Henrissat, B., "Detection of secondary structure elements in proteins by hydrophobic cluster analysis", *Prot. Eng.* **5**, 629–635 (1992).

22. Bourat, G., Thoreau, E. and Mornon, J.P., "2-D helical hydrophobic clusters: Statistics and morphology", *J. Pharm. Belg.* **49**, 226–235 (1994).

23. Callebaut, I., Tasso, A., Brasseur, R., Portetelle, D., Burny, A. and Mornon, J.P., "Common prevalence of alanine and glycine in reactive center loops of serpins and viral fusion peptides: Do prions possess a fusion peptide?" *J. Comp. Aided Mol. Design* **8**, 175–191 (1994).

24. Forloni, G., Angeretti, N., Chiesa, R., Monzani, E., Salmona, M., Bugiani, O. and Tagliavini, F., "Neurotoxicity of a prion protein fragment", *Nature* **362**, 543–546 (1993).

Glossary

α-helix: regular helical path of the polypeptide chain (about 6-20 residues).

β-sheet: association of several (2-10) β strands in a parallel, antiparallel or mixed way and linked by H bonds.

β-strand: extended regular "zig-zag" portion of the polypeptide chain (about 3-10 residues).

cytokine receptors: family of membraneous receptors mediating the action of numerous protein hormones such as the interleukines, the interferons, the growth hormones, etc.

codon: Set of three consecutive nucleotides along the DNA currently coding for an amino acid.

HCA score: numerical score reflecting the similarity of hydrophobic clusters in HCA plots.

homology score: numerical index characterizing the similarity between sequences, e.g., sequence identity score expressed as a percentage.

HSC70: heat shock cognate protein of 70 kilodalton.

human CD4: receptor on the surface of T cells which is a main target for the HIV virus (AIDS).

hydrophobic cluster: group of hydrophobic amino acids which appear in the helical representation of sequences, its counterpart in the 3-D structure being generally included in a regular *secondary structure*: (β strand or helix).

laminin B2: A chain of a large basement membrane glycoprotein involved in various biological activities such as cell adhesion, migration, differentiation, etc. . .

open reading frame (ORF): part of the DNA chain which leads to the translation of the genetic code into a polypeptidic chain (protein).

protein sequence or protein primary structure: arrangement of the 20 (currently) possible amino acids along a polypeptide chain.

secondary structure: local 3-D organization of a polypeptide chain forming regular structures (β strands and helices) or mainly irregular ones (loops).

sequence identity level: percentage of identical amino acids between two aligned sequences.

stop codon: codon which interrupts the translation into a polypeptidic chain of a gene along the DNA chain. Stop codons occur at the end of an open reading frame (ORF).

tertiary structure: 3-D folding pattern of the polypeptide chain.

turn: a particularly short loop.

2-D helical net: 2-D display of sequence as the projection of a helix (cf., Fig. 2).

Diagnosis of Complex Patterns in Protein Sequences

T.K. Attwood
and
D.J. Parry-Smith

A system for diagnosing complex patterns within protein sequences is described. The method uses composite pattern-recognition discriminators to scan individual query sequences, and displays the results graphically, giving an instant evaluation of diagnostic performance in that sequence. The graphs yield separate profiles for each element of the composite discriminator, each with a statistical breakdown of the result, thereby providing both qualitative and quantitative assessments of discriminating power. The method is useful for the purpose of confirmation, in sequences whose family membership is already known, or of profile verification, in sequences whose family membership is only suspected. Since each element of the discriminator is treated independently, the graphs may also indicate whether a sequence is characterised by only part of a particular pattern, or indeed by none of it. Examples are provided to illustrate the characteristic signature of the superfamily of G-protein-coupled receptors, and those of kringle domains, zinc fingers and complement-binding domains.

Introduction

With the recent revolution in DNA sequencing techniques, the rate of acquisition of protein sequence data (through translation of sequenced DNA) has entered an exponential phase. It is clear that hidden within this vast pool of amino acid sequences, which have been collected together in protein sequence databases, lies a wealth of structural and functional information. In the past, we have relied almost exclusively on crystallographic techniques to learn about the 3-D structures of proteins, but these methods can be immensely time-consuming and cannot keep pace with the explosive rate at which protein sequences are now being deduced. As a result, methods are now being sought to analyse the primary amino acid sequences themselves in order to gain a greater insight into the relationship between sequences and their structures and functions.

One of the most important steps in the analysis of protein sequences is the process of multiple sequence alignment. This involves lining up the linear amino acid sequences of the same or similar proteins from different species in such a way that corresponding parts of the sequences are brought into equivalent positions — very often, this procedure involves the insertion of gaps between particular residues to bring adjacent sequences into the correct juxtaposition. As more sequences are added to an alignment in this way, patterns tend to emerge whereby insertions are confined to specific regions that flank conserved ungapped blocks of the alignment. Closer inspection reveals that these conserved blocks often relate to core structural elements of the protein, such as the helical or sheet components of the structure, while the gapped regions are more likely to occur in the linking loop regions between core elements.

Sequence alignments may therefore hold vital structural and functional clues within the patterns of conservation they reveal. These conserved regions are often referred to diversely as motifs, features, patterns, signatures, modules, or simply domains. We generally use the term motif to describe a single conserved structural or functional region and groups of motifs, which together provide a signature or fingerprint for a particular family of sequences, are then referred to as patterns. A complex pattern is one in which a simple pattern is repeated many times along the length of a sequence.

Extrication of the structural and functional information latent in protein sequences is the goal of sequence analysis. Consequently, there are now numerous methods available to search for either characteristic motifs or patterns in large sequence databases (e.g., references 1, 2, 5–10, 12–15). Our method differs in various ways from all of these and uses composite discriminators to search for particular patterns either in sequence databases or in single query sequences. A discriminator is the actual set of equivalent aligned sequence motifs excised from a conserved region of a multiple alignment — the scanning algorithm interprets these aligned sequence fragments in terms of residue frequencies at each position in the aligned fragment. A composite discriminator is then the set of discriminators that together encode a pattern. The relationship between motifs, patterns and discriminators is illustrated in Fig. 1.

To demonstrate the presence of patterns and complex patterns within protein sequences, we use plots to portray matches between a given query sequence

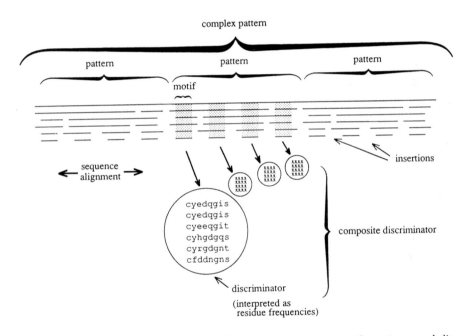

Fig. 1. Schematic diagram illustrating the relationship between motifs, patterns and discriminators. The broken parallel lines represent a set of aligned protein sequences, with the most similar sequences at the top of the alignment and the least similar sequences towards the bottom — gaps indicate points at which insertions have been made to bring equivalent parts of the sequences into the correct register. Shaded regions, where the alignment is preserved, are termed motifs, groups of motifs forming discrete patterns (the same pattern repeated a number of times in a sequence is referred to as a complex pattern). The excerpts denote excised, aligned sequence fragments that are used to form a discriminator — the scanning algorithm interprets these as residue frequencies at each position of the fragment. Groups of discriminators, which together describe discrete patterns, are termed composite discriminators.

and each individual element of the pattern. These plots provide excellent diagnostic tools, and can be used to differentiate between those sequences that are characterised by the whole of a particular pattern, those that contain only part of it, and those that contain none of it. For the purpose of illustration, we have chosen to display the patterns encoded by four composite discriminators, namely, those for: (a) the superfamily of G-protein-coupled receptors (GPCR's) — these are membrane proteins involved in signal transduction, and are characterised by structures that are believed to comprise 7 transmembrane helices; (b) kringle domains — these are triple-looped, disulphide cross-linked structures found in varying numbers of copies in some serine proteases and plasma proteins; (c) zinc fingers — these are nucleic acid-binding structures comprising ≈30 residues, including cysteine and histidine residues in tetrahedral coordination with a zinc atom; and (d) complement-binding domains — these are cell adhesion membrane proteins that bind components of the complement system (this is a complex group of proteins in the blood that, working together with antibodies or other factors, plays an important role as a mediator of immune and allergic reactions). The sequences selected for scanning include representative examples both of known true family members and of non family members.

Method

The graphical method used here takes each of the excised aligned motifs that comprise a composite discriminator and uses them in turn to search a query sequence. The result is shown as a collation of plots or windows, one for each element of the discriminator. The horizontal axis of the plot depicts the sequence itself, and the vertical axis the percentage score of each element of the composite discriminator (0–100 per element). A match with any element of the discriminator appears as a sharp peak above the background noise — the peak denotes a residue by residue match with the sequence, its leading edge marking the first position of the match. For every window, a statistical analysis of the highest peak is provided in order to give a measure of confidence for that match.

Algorithm

The program uses a linear, sliding window algorithm for calculating a score at each residue position in the sequence. Each motif is scanned across the sequence independently of the others. Specifically, the sequence is broken up into sections equal to the length of the motif concerned. Each section overlaps the last by (motif_length-1) residues; a score for the last (motif_length-1) residues in the sequence cannot be calculated. The motif is translated into a table of residue frequencies and the maximum score that a section could achieve is calculated by summing the highest frequency value in each motif position and then multiplying this by the length of the motif. A score is calculated for each section by looking up successive residue positions in the table, and the resulting frequencies are summed. Where a position in the current section achieves a score greater than 0, a counter for the section is incremented. In calculating the percent score, the sum for the section is multiplied by the counter; this has the effect of enhancing the signal to noise ratio. The score for each residue in the sequence is plotted in a window, which is always scaled between 0 and 100 percent in order to allow comparison between plots.

The program is written in VAX C and uses VAX GKS (C binding) for its graphical output, hence supporting a wide range of display devices; it runs interactively. In its present form, the program has been incorporated into a larger software package for sequence analysis, known as ADSP,[11] which is itself part of the SERPENT integrated database and software system,[3] currently available through the SEQNET service at Daresbury, UK.

Results

The familiar signature of the GPCR's is a characteristic pattern of 7 membrane-spanning regions. The composite discriminator for this family of receptors therefore comprises 7 elements, 1 for each of the transmembrane motifs.[16] The plots in Fig. 2 illustrate the results of searching 4 different sequences with the GPCR composite discriminator: (a) shows 7 distinct peaks (all measuring between 8 and 15 standard deviations (σ) above the mean), corresponding to each of the transmembrane regions, indicating that each membrane-spanning region can be identified by its own characteristic motif and that the sequence is therefore a true family member — note that the sequence is characterised by a large loop region between the fifth and sixth motifs; (b) depicts 7 less well-defined peaks, each of which appears in a systematic order above the level of noise (between 5 and 10 σ above the mean), so this too is a member of the true set, if a little more distantly related — this sequence does not possess the large cytoplasmic loop seen in (a); (c) shows only 6 reasonably strong peaks

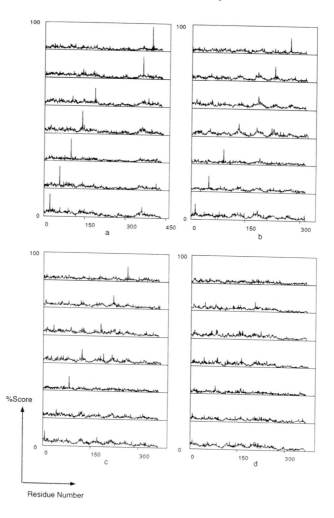

Fig. 2. Composite discriminator scans for the superfamily of G-protein-coupled receptors: The discriminator comprises 7 elements, one for each of the membrane-spanning motifs. The plots shown are those for (a) the porcine muscarinic acetylcholine receptor; (b) the guinea pig platelet activating factor receptor; (c) the human cytomegalovirus GPCR homologue UL33; and (d) the *Dictyostelium discoideum* cyclic-AMP receptor. The first 3 plots depict true members of the superfamily, but the third shows only a partial match with the discriminator, perhaps pointing to a subfamily lacking critically conserved residues in the second transmembrane motif. The final plot illustrates a sequence that does not appear to be a member of the superfamily, where no peak or set of peaks makes a significant impact above the level noise (the highest peaks measure between 5 and 15 σ above the mean in plots (a) to (c), and below 5 σ above the mean in (d)).

(each 5 to 11 σ above the mean), indicating that while this sequence is closely similar to the superfamily, it makes only a poor match with the second motif (4.5 σ above the mean); and finally, (d) displays no peak or set of peaks in a systematic order above the level of noise (all measure below 5 σ above the mean), showing that the GPCR discriminator does not diagnose the sequence of this membrane protein as a member of the GPCR superfamily. Note that the method does not imply that this last sequence is not a G-protein-coupled receptor — it simply illustrates the point that, at the sequence level, it cannot be identified with the rest of the superfamily.

By contrast with the relatively simple situation represented by the GPCR's, Fig. 3 illustrates the use of composite discriminators in the diagnosis of rather more complex patterns: (a) depicts a search with a 4-element composite discriminator for kringle domains (K.M. Measures, unpublished data) — the results reveal that the sequence under investigation makes multiple matches with

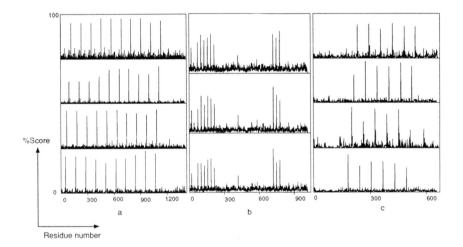

Fig. 3. Complex composite discriminator scans showing (a) kringle domains in the rhesus macaque apolipoprotein A sequence; (b) zinc fingers in the mouse ecotropic virus integration 1 site protein; and (c) complement binding domains in the mouse endothelial leukocyte adhesion molecule 1 precursor. The plots depict multiple occurrences of the patterns in question: The kringle domain contains 4 separate motifs, and the sequence searched is characterised by 10 consecutive kringle repeats; the zinc finger pattern comprises 3 elements, and the query sequence contains 11 repeats unequally shared between 3 distinct regions; and the complement binding domain has 4 separate elements, the target sequence containing 6 regularly spaced repeats. In each case, the highest peaks register 9 to 15 σ above the mean.

the discriminator, in this case pinpointing 10 consecutive repeats at more or less regular 120 residue intervals from the beginning of the sequence; (b) portrays a scan with a 3-element composite discriminator for zinc fingers (M. Beck, unpublished data) — the result again shows multiple matches, but further than this it clearly illustrates that the 11 repeats, which for the most part occur at 25 residue intervals, are unevenly distributed along the sequence, clustering in 3 discrete regions; and finally, (c) shows a search with a 4-element composite discriminator for complement-binding domains — once more, multiple repeats are apparent, but here there are only 6 occurrences of the pattern, situated towards the middle of the sequence and spaced approximately 60 residues apart. In all of these cases, the strongest peaks measure between 9 and 15 σ above the mean.

Discussion

The results demonstrate that a simple graphical approach can be used very effectively to illustrate fine details of characteristic sequence signatures. In particular, it can be used to show subtle differences between known family members (including disparate motif separations, different qualities of motif matches, etc.), it can be used to highlight sequences that are characterised by only part of a pattern, or it can pinpoint sequences that are not matched by the pattern at all. Further, the plots can reveal multiple pattern repeats and indicate whether the repeats occur in regular arrays, whether they are confined to specific regions in the sequence, and so on. Thus patterns of varying complexity can be diagnosed with the system.

Numerous pattern-recognition methods now exist to search for characteristic motifs. Some of these attempt to encode complete protein folds by creating patterns of core structural domains, and matches are *only* recognised if they contain all the constituent elements of the pattern (e.g., reference 5); others generate so-called profiles using groups of aligned sequences, where the alignment may be the full sequence length, or just a single structural domain (e.g., reference 9). Indeed, the relative simplicity of dealing with single domains, rather than complete protein folds, has led to a proliferation of techniques that concentrate on defining unique structural or functional motifs (e.g., references 1, 2, 12, and 13).

Our method combines attributes of a number of these approaches. A crucial difference, however, lies in the fact that we always define a pattern in terms of a number of motifs and in all such instances we treat each element of the pattern *independently*. The use of more than one motif in itself greatly enhances the probability of successful recognition of a pattern, because each element contributes to the formation of a characteristic 'fingerprint'. The independent treatment of each element has the further advantage that investigation of composite patterns is made very simple and, more importantly, potentially valuable information about possible subfamilies is not lost because *partial* matches with the pattern are not disregarded. In the case of the G-protein-coupled receptors, for example, the most important diagnostic feature is not the specific height of individual peaks but, exactly as with a fingerprint, it is the pattern they make as a whole: This is why we accept the human cytomegalovirus receptor as a true family member, even though one of the 7 matches is very weak (it perhaps belongs to a subfamily in which critically conserved residues in the second transmembrane motif are absent); but we dismiss the cAMP receptor, because what peaks exist are very small and do not appear to make contributions in a systematic way — in other words, there is no discernible pattern.

Composite discriminator plots of the type shown here can highlight the presence, and illustrate the nature, of a variety of complex composite patterns. The appearance of the plots can be both striking and informative, and as such can be invaluable as an immediate and effective diagnostic tool.

Acknowledgements

We wish to thank Kathryn Measures and Michael Beck for generously providing us with their discriminators for kringles and zinc fingers.

References

1. Abarbanel, R.M., Wieneke, P.R., Mansfield, E., Jaffe, D.A. and Brutlag, D.L., "Rapid searches for complex patterns in biological molecules", *Nucleic Acids Res.* **12**, 263–280 (1984).
2. Akrigg, D., Bleasby, A.J., Dix, N.I.M., Findlay, J.B.C., North, A.C.T., Parry-Smith, D.J., Wootton, J.C., Blundell, T.L., Gardner, S.P., Hayes, F., Islam, S., Sternberg, M.J.E., Thornton, J.E., Tickle, I.J. and Murray-Rust, P., "A protein sequence/structure database", *Nature* **335**, 745–746 (1988).
3. Akrigg, D., Attwood, T.K., Bleasby, A.J., Findlay, J.B.C., Maughan, N.A., North, A.C.T., Parry-Smith, D.J. and Perkins, D.N., "SERPENT: An information storage and analysis package for protein sequences", *CABIOS* **8**(5), 451–459 (1992).

4. Attwood, T.K. and Findlay, J.B.C., "A potent discriminator for the superfamily of G-protein-coupled receptors: Use of a compound feature index", *Protein Engineering* **6**(2), 167–176 (1993).

5. Barton, G.J. and Sternberg, M.J.E., "Flexible protein sequence patterns. A sensitive method to detect weak structural similarities", *J. Mol. Biol.* **212**, 389–402 (1990).

6. Bowie, J.U., Luthy, R. and Eisenberg, D., "A method to identify protein sequences that fold into a known three-dimensional structure", *Science* **253**, 164–170 (1991).

7. Cockwell, K.Y. and Giles, I.G., "Software tools for motif and pattern scanning: Program descriptions including a universal sequence reading algorithm", *CABIOS* **5**(3), 227–232 (1989).

8. Devereux, J., Haeberi, P. and Smithies, O., "A comprehensive set of sequence analysis programs for the VAX", *Nucleic Acids Res.* **12**, 387–395 (1984).

9. Gribskov, M., McLachlan, A.D. and Eisenberg, D., "Profile analysis: Detection of distantly related proteins", *Proc. Natl. Acad. Sci. USA.* **84**, 4355–4358 (1987).

10. Lathrop, R.H., Webster, T.A. and Smith, T.F., "Ariadne: Pattern-directed inference and hierarchical abstraction in protein structure recognition", *Communications of the ACM* **30**, 909–921 (1987).

11. Parry-Smith, D.J. and Attwood, T.K., "ADSP: A new package for computational sequence analysis" *CABIOS* **8**(5), 451–459 (1992).

12. Sibbald, P.R. and Argos, P.A., "Scrutineer: A computer program that flexibly seeks and describes motifs and profiles in protein sequence databases", *CABIOS* **6**(3), 279–288 (1990).

13. Smith, H.O., Annau, T.M. and Chandrasegaran, S., "Finding sequence motifs in groups of functionally related proteins", *Proc. Nat. Acad. Sci. USA* **87**, 826–830 (1990).

14. Staden, R., "Methods to define and locate patterns of motifs in sequences", *CABIOS* **4**(1), 53–60 (1988).

15. Taylor, W.R., "Identification of protein sequence homology by consensus template alignment", *J. Mol. Biol.* **188**, 233–258 (1986).

General References

16. Attwood, T.K. and Findlay, J.B.C., "Multiple sequence alignment of protein families showing low sequence homology: A methodological approach using database pattern-matching discriminators for G-protein-coupled receptors", *Gene* **98**, 153–159 (1991).

17. Cabot, E.L. and Beckenbach, A.T., "Simultaneous editing of multiple nucleic acid and protein sequences with ESEE. *CABIOS* **5**(3), 233–234 (1989).

18. Devereux, J., Haeberi, P. and Smithies, O., "A comprehensive set of sequence analysis programs for the VAX", *Nuclei Acids Res.* **21**, 387–395 (1984).

19. Faulkner, D.V. and Jurka, J., "Multiple aligned sequence editor (MASE)", *TIBBS* **13**, 321–322 (1988).

20. Findlay, J.B.C., Eliopoulos, E.E. and Attwood, T.K., "The structure of G-protein-linked receptors", *Biochem. Soc. Symposia* **56**; *G-Proteins and Signal Transduction*, eds. G. Mulligan, M.J.O. Wakelam and J. Kay (1990) pp. 1–8.

21. Moereels, H., De Bie, L. and Tollenaere, J.P., "CGEMA and VGAP: A colour graphics editor for multiple alignment using a variable GAP penalty. Application to the muscarinic acetylcholine receptor", *J. of Computer-Aided Mol. Design* **4**, 131–145 (1990).

22. Morris, G.M., "The matching of protein sequences using color intrasequence homology displays", *J. Mol. Graphics* **6**, 136–140 (1988).

23. Parry-Smith, D.J and Attwood, T.K., "SOMAP: A novel interactive approach to multiple protein sequence alignment", *CABIOS* **7**(2), 233–235 (1991).

24. Staden, R., "Methods to define and locate patterns of motifs in sequences", *CABIOS* **4**(1), 53–60 (1988).

25. Stockwell, P.A. and Petersen, G.B., "HOMED: A homologous sequence editor", *CABIOS* **3**(1), 37–43 (1987).

Glossary

multiple alignment: A linear comparison of amino acid sequences in which insertions are made in order to bring equivalent positions in adjacent sequences into the correct register. Alignments are used in particular to pinpoint the occurrence of characteristic motifs that have been preserved in all sequences in the alignment.

motif: Any consecutive string of amino acids in a protein sequence whose general character is repeated, or conserved, in all sequences in a multiple alignment at a particular position. Motifs are of interest because they may correspond to structural or functional elements within the sequences they characterise.

pattern: A set of motifs within a protein sequence alignment that together provide a signature for a particular family of sequences. A pattern may represent, for example, a group of structural elements, such as a set of α helices, a group of β strands, an assembly of strands and helices, and so on.

complex pattern: A simple pattern that is repeated any number of times along the length of a protein sequence. Proteins that are characterised by multiple repeats tend to be involved in protein-protein interactions or protein-DNA interactions (e.g., such as complement-binding and kringle domains, zinc fingers, etc.).

discriminator: The set of equivalent aligned motifs excised from a given position in a sequence alignment and used to search either a given query protein sequence or a database for the occurrence of the same or similar motif.

composite discriminator: The set of discriminators that together encode a pattern within a protein sequence alignment, and used to search either a given query sequence or a database for the occurrence of the same or similar pattern.

discriminating power or diagnostic performance: A measure of the ability of a discriminator or composite discriminator to identify true matches, either in a database or in an individual query sequence.

true set: The family of sequences in a database known to possess a particular motif or pattern. In this latter context, the true set is only that group of sequences possessing all elements of the pattern.

G-protein-coupled receptors: Membrane-bound receptors linked to guanine nucleotide binding proteins (G-proteins). They represent a diverse family of signal transduction mediators, whose activating ligands vary widely in structure and character. Members of the family include: the visual and olfactory receptors; muscarinic acetylcholine, β adrenergic and dopaminergic receptors; and so on. The proteins appear faithfully to have conserved a basic structural framework consisting of 7 transmembrane helices, whose packing arrangement has been modeled on that of bacteriorhodopsin.

kringle domain: A small triple-looped, disulphide cross-linked domain found in varying numbers of copies in some serine proteases and plasma proteins. Kringles have been found in apolipoproteins, blood coagulation factors, plasminogens, thrombins and so on. They are believed to play a role in binding mediators such as membranes, other proteins, or phospholipids, and in the regulation of proteolytic activity.

zinc finger: A nucleic acid-binding structure comprising around 28 residues, including 2 conserved cysteine residues and 2 conserved histidines in tetrahedral coordination with a zinc atom. The 12 residues between the second cysteine and the first histidine within this C2H2 repeat are largely basic and polar, implicating this region in particular in nucleic acid binding. The C2H2 type zinc fingers, which tend to occur as multiple tandem repeats, are one of a number of different types of zinc finger domain.

complement-binding protein: Membrane proteins that bind components of the complement system. They include, for example, leukocyte adhesion molecules, lymph node homing receptors, granule membrane proteins, and so on. The complement-binding domain may be present in varying numbers of copies within each of these different proteins.

complement system: A complex group of proteins in the blood that acts on its own and in cooperation with antibodies in defending vertebrates against infection. The proteins are activated sequentially in an amplifying series of reactions either by the classical pathway (triggered by IgG or IgM antibodies binding to antigen), or by the alternative pathway (triggered by the cell envelopes of invading microorganisms).

RNA Folding and Evolution

Kenji Yamamoto

and

Hiroshi Yoshikura

Existing genes are compared with random nucleotide sequences of the same base composition for their ability to form stem-loop structures. RNA viral sequences are more "folded" than the random sequences, while the prokaryotic transcripts are more "extended". In the eukaryotic transcripts, the coding regions are more "neutral" in folding than the non-coding regions.

The 3-D folded structure of RNA has evolved throughout time and affects various functions of mRNA, tRNA, snRNA or rRNA. This property of RNA must have evolved as other biological parameters have.

We previously devised a computer algorithm predicting the folding structure of a single stranded RNA. The predicted structures of tRNA, snRNA and viroids agreed well with the experimentally confirmed ones.[1-3]

To express the degree of global folding of RNA, we therefore derived a value called information mass volume (IMV).[2] IMV is defined as the sum total of the information values[1,4] reflecting the degree of the folding of a given sequence. Generally, if a sequence contains more regions of possible intra-strand annealing, the IMV value is higher.

Hardware, Software and Database

The computer systems used are the VAX8530 (VMS 5.0) and VAXSTATION 3100 (VMS Version 5.2),[5] and the programs are written in FORTRAN.

The databases used are GENBANK[6] (Release 68) and EMBL[7] (Release 27). The genes used for the analysis are listed in Table 1. The GCG

Table 1.

Genes		Normalized IMV	
		Original	Random
Virus			
Turnip yellow mosaic virus coat protein cds		105	4
Poliovirus type 1 complete genome		360	5
Moloney murine leukemia virus complete genome		460	85
Bacteriophage MS2 complete genome		189	57
Sindbis virus capsid protein cds		378	45
Sindbis virus glyco-protein cds		89	7
Semliki forest virus complete genome		405	205
Foot and mouth disease virus complete genome		224	105
Retrovirus NK24 v-fos gene cds		839	224
Reticuloendotheliosis virus pol and env gene		101	20
Hepatitis E virus polyprotein mRNA		399	5
Avian cartinoma virus MH2 gag v-mil and v-myc genes		35	9
Bovine enterovirus VG5-27 complete genome		33	32
Bovine leukemia virus LB59 env cds		55	8
La crosse virus S RNA cds		11	1
Bovine leukemia virus gag and pol genes		651	60
Encephalomyocarditis virus EMC-B complete genome		52	2
Dengue virus type 2 complete genome		9	2
Human imunodeficiency virus type 2 complete genome		50	5
Feline imunodeficiency virus complete genome		43	1
Human coronavirus 229 genes mRNA		8	1
Hepatitis A virus complete genome		54	1
Japanese encephalitis virus complete genome		97	25
Cucumber mosaic virus RNA3		124	14
Vesicular somatitis virus glyco protein cds		39	7
Vesicular somatitis virus nucleo protein cds		41	45
Rabies virus complete genome		35	4
Tabacco mosaic virus complete genome		20	2
Measles virus complete genome		204	125
Sendai virus M protein cds		53	13
Human para-influenza virus HN mRNA		27	1
Influenza seg6 cds		10	5
Influenza seg8 cds		87	2
Influenza seg1 cds		7	3
Human corona virus 229 genes mRNA		8	1
Lymphocytic choriomeningitis virus S RNA cds		78	10
Lymphocytic choriomeningitis virus RNA polymerase cds		10	6
New castle disease virus HN RNA cds		48	8
Abelson v-abl phosphotyrosine acceptor site		16	10
Feline leukemia virus v-myc		238	202
Reticuloendotheliosis virus v-rel		82	52
Avian sarcoma virus src		324	25
Prokaryotes			
E. coli recA		173	228
E. coli lactose permease	(lacY)	8	5
E. coli beta-galactosidase	(lacZ)	16	24
E. coli tryptophan synthetase	(trpE)	88	256
E. coli tryptophan synthetase	(trpD)	75	247
E. coli tryptophan synthetase	(trpC)	7	198
E. coli tryptophan synthetase	(trpB)	51	567
E. coli tryptophan synthetase	(trpA)	3	189
E. coli tryptophan synthetase	(trpE)	197	260
E. coli peptidase D	(pepD)	30	108
E. coli soxR and soxS genes		3	2
E. coli ATP synthetase	(unc)	74	90
E. coli DNA glycosylase I	(tag)	8	62

Table 1. (*Continued*)

E. coli enterotoxin	(estA4)	5	105
E. coli stavation protein	(SSP)	33	62
E. coli single-strand DNA binding protein (ssb)		77	119
L. lactis citrate permease (citP)		8	8
M. capsulatus glutamin synthetase (glnA)		475	500
K. pneumoniae nitrogen regulation (ntrB)		127	101
N. gonorrheae penicillin binding protein (penA)		53	92
S. typhimurium alanine ractamase (air)		243	263

Eukaryotes

Yeast (S. cerevisiae) pyruvate decarboxylase (PDC1)		27	4
Yeast (S. cerevisiae) mating type genes		5	4
D. melanogaster antennapedia (antp)		6	650
D. melanogaster heat shock protein (hsp70)		0	210
D. melanogaster calmodulin gene		2	55
D. melanogaster virilis sevenless genes		121	11
D. melanogaster heteroneura alchol dehydrogenase		38	1
D. melanogaster cuticle protein		168	10
D. melanogaster bicoid gene (bcd)		125	3
Soybean actin gene		43	56
Tomato fruit ripening gene		8	2
Angelfish preproinsulin		118	87
Xenopus laebis type I keratin		192	22
Chiken ovalmin		8	438
Chikin histon h2b		1850	120
Chiken pro-alpha2 collagen		15	201
Chiken myosin alkali 1-chain		8	120
Chiken brain tubulin beta chain		85	176
Mouse pancreatic alpha-amylase		25	3
Mouse transplantation antigen		306	820
Rabbit beta 1 globin type 1		45	78
Human enkephalin		173	110
Human hla-dr antigen heavy chain		78	8
Human leukocyte interferon		2	98
Human hla-dr antigen beta chain		789	1056
Human hla-dr antigen alpha chain		15	125
Human humos transforming gene		1006	1501
Human cmos transforming gene		985	1105
Human ig kappa 1 chain constant region		30	205
Human histone H1		3	2

software package was used for the database handling. Random sequences were obtained by the random number generating program in VAX11/780 RUNTIME LIBRARY. The information mass volume (IMV) for a given base ratio did not fluctuate widely among different randomization.

The derivation of IMV is outlined as follows. It consists of two steps.

[1] All the possible stem-loops in a given RNA are generated in the computer. They are numbered $S_1, S_2, \ldots, S_n$. The free energy level of each stem-loop is calculated. The position, the size, and the free energy of each stem-loop are stored in the disk memory (STEM-LOOP DATA). Then, all the structures with the compatible stem-loops are generated using the STEM-LOOP DATA.

[2] The possible frequency of each stem-loop occurrence was estimated as follows. For a given stem-loop S_i, all the possible folding structures which

contain the stem-loop S_i are generated in the computer with reference to the STEM-LOOP DATA using the EIGHT QUEEN METHOD (1). The free energy of each structure, $e_{i1}, e_{i2}, \ldots, e_{im}$ (m is the number of the structures generated) is calculated. Here, we define the frequency of a stem-loop S_i as

$$q_i = \sum_{j=1}^{m} \log(e_{ij}).$$

The IMV is defined as

$$\text{IMV} = \sum_{i=1}^{n} \log(q_i).$$

For comparison of sequences with various sizes, the IMV was normalized for 20 bases, i.e.,

Normalized IMV = IMV $\times$ (20/nucleotide length of the gene).

The flow chart of the algorithm has been already published.[2] The program is available through K. Yamamoto.

Results

The normalized IMV was calculated for each gene and for random sequences with the same base composition. The normalized IMV of a given gene is plotted along the horizontal axis (X) against that of the corresponding random sequence in the vertical axis (Y).

Figure 1(a) shows such a plot for single strand prokaryotic RNA virus genomes and eukaryotic RNA virus genomes. All the plots are below the line $Y = X$, indicating that the actual genes are more "folded" than the corresponding random sequences.

Figure 1(b) shows a plot for various chromosomal transcripts. The eukaryotic transcripts (open circles) are distributed on both sides of the line $Y = X$, while all the prokaryotic mRNA (closed circles) are above the line.

The eukaryotic transcripts have long sequences of 5' and 3' non-coding regions and introns, while prokaryotic mRNA has negligibly short non-coding stretches. The plots for the whole transcript and those for the coding region were compared for the eukaryotic genes. The plots both below and above the line $Y = X$ became closer to the line after removal of the non-coding region, i,e., the folding level of coding regions are close to that of the random sequence of the same ratio. Here, we call the coding regions "neutral" in folding.

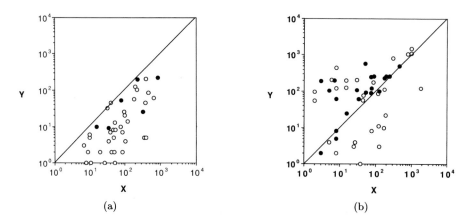

(a)   (b)

Fig. 1. IMV plot for actual gene sequences versus random sequences with the same base ratio. The normalized IMV for actual genes is plotted along the X axis: Normalized IMV for the corresponding random sequences is plotted along the Y axis. (a) Single strand RNA viruses (open circles) and naturally transduced oncogenes (filled circles). (b) Chromosomal transcripts of eukaryotes (open circles) and prokaryotes (filled circles).

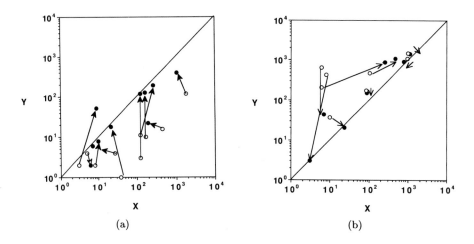

(a)   (b)

Fig. 2. Plot of normalized IMV for eukaryotic transcripts. Primary transcripts (open circles) are related to their coding regions (filled circles) by arrow. (a) Genes whose primary transcripts are more "folded" than the random sequences. (b) Genes whose transcripts are more "extended" than the random sequence.

Speculations

The fact that viral sequences are more "folded" than the random sequences is compatible with the fact that viral genes have to be packaged into virions. The host derived oncogenes so for examined (closed circles in Fig. 1(a)) were also below the line $Y = X$. This may suggest that there is a conformational restraint on the transcript of the genes naturally transduced. The more "unfolded" structure of prokaryotic genes will be evolutionary advantageous in that it allows rapid translation and the shorter mRNA half both actually seen in the prokaryotes.

In eukaryotes, with few exceptions, the coding regions are more "neutral" in folding than the whole transcripts. This fits well with the fact that the function of the non-coding region resides in the structure itself while that of the coding region primarily resides in the translated amino acid sequence and not in the secondary structure of RNA.

Existing genes from different kingdoms show clearly different patterns for their probability of forming stem-loop structures, and these probabilities are different from those of random sequences of the same base composition. The evolutionary implications of these relationships is fertile ground for future research.

Acknowledgement

We thank for Ms. Kobayashi for preparing the manuscript.

References

1. Yamamoto, K., Kitamura, Y. and Yoshikura, H., "Computation of statistical secondary structure of nucleic acids", *Nucl. Acids Res.* **12**(1), 335–346 (1984).
2. Yamamoto, K. and Yoshikura, H., "Relation between genomic and capsid structures in RNA viruses", *Nucl. Acids Res.* **14**(1), 389–396 (1986).
3. Yamamoto, K., Sakurai, N. and Yoshikura, H., "Graphics of RNA secondary structure: Towards an object-oriented algorithm", *CABIOS* **3**(2), 99–103 (1987).
4. Yamamoto, K. and Yoshikura, H., "Computer programme statistically predicting folding structure of nucleic acids of any length and example of calculation", *Jpn. J. Exp. Med.* **54**(6) 241–247 (1984).
5. Digitial Equipment Corporation, Maynard, Massachusetts, USA.

6. Bilofsky, H.S., Burks, C., Fickett, J.W., Goad, W.B., Lewitter, F.I., Rindone, W.P., Swindell, C.D. and Tung, C.S., "The GenBank genetic sequence databank", *Nucl. Acids Res.* **14**(1), 1–4 (1986).
7. Hamm, G.H. and Cameron, G.N., "The EMBL data library", *Nucl. Acids Res.* **14**(1), 5–10 (1986).

Glossary

prokaryote: An organism lacking a nuclear membrane and certain organelles such as mitochondria.

eukaryote: An organism with a nuclear membrane and certain organelles such as mitochondria and mitotic spindle.

snRNA: (Small nuclear RNA) is any one of many small RNA species confined to the nucleus; several of the snRNAs are involved in splicing or the RNA processing reactions.

rRNA: (Ribosomal RNA) is one of the component of ribosome.

GENBANK: One of the genetic sequence data banks.[5]

EMBL: One of the genetic sequence data banks.[6]

GCG: Sequence analysis software package by Genetics Computer Group, Inc., University Research Park, 575 Science Drive, Suite B, Madison, Wisconsin 53711, Phone: (608) 231-5200, Fax: (608)231-5202.

Representation of Biological Sequences Using Point Geometry Analysis

Y.K. Huen

In the past decade, researchers such as Hamori,[2-4] Gates,[5] and Pickover[6] have mapped genetic letter-series sequences using arbitrary vector-codes. These codes are based on vectors in either Cartesian 2- or 3-space. Recently several papers were published on the development of point geometry which is not based on the Cartesian system of coordinates but on the multidimensional Hgram system of coordinates.[1,9-16] Using the Hgram graph as a reference, a method of classifying graphs is described. A list of desirable design criteria for vector-codes is given. The design procedure of Huen's vector-code for protein sequences is outlined. Included is the outline of the specifications for writing software for the visual representation of protein sequences on microcomputers. The new code has some advantages over previous codes in that there are no overlaps of individual vectors and no requirement for spatial rotational viewing. Each basic symbol can embody a large amount of information, and only 2-D graphical software is needed for its implementation. It is suggested that letter-series sequences should be viewed as points in n-dimensional space. Visual analyses using the new code is advantageous since the mapping can be done without any loss of information from the IUB standard symbol table.

Introduction

Genetic sequences are conventionally expressed in letter-series formats since these are easily manipulated and analyzed by computers, and compactly stored as ascii-files in magnetic media. Unfortunately, for long sequences of over a million nucleotide bases, pattern recognition using letter-series sequences is ineffective since the human eyes are more adept at recognising 2-D patterns than 1-D text-strings. This led researchers such as Hamori,[2–4] Gates[5] and Pickover[6] to investigate computer-aided visualization of DNA sequences using arbitrary vector-codes. These are arbitrary because the patterns generated do not reflect the stereochemical structures of the genomes.

It is interesting to observe that in this 3-D world of ours, information is stored in 1-D and 2-D form. When we take a snapshot of a beautiful scene, we are mapping points from the 3-D space into a 2-D photographic film. For convenience, we may call this a 3:2 mapping. We know for sure that in this mapping, we have lost one spatial dimension which results in an absence of depth of field in the photograph, the target space. Mathematically we say that the mapping of points is non-bijective since million of points along a ray of light are mapped to a single point in the photograph. Our commonest display media, such as the CRT-screen and the printed page, are limited to two dimensions. If the Cartesian system of coordinates is used, then only a 2:2 mapping is bijective. A classical example is the casting of the shadow of a skeletal cube by parallel light onto a flat screen whereby the locations of the diagonally opposite vertices are visually ambiguous due to a loss of one spatial dimension in the 3:2 projection.

This chapter describes a novel multidimensional graphical format based on the Hgram system of coordinates.[1,9–17] "Hgram" is the acronym for Huen's diagram. Just as Euclid's geometry is based on the Cartesian system of coordinates, a new geometry called point geometry is based on the Hgram system of coordinates. The axiom, theorems, and applications in several areas of graphical analyses of real world problems have been described in detail in papers listed in the reference section. Much of this body of geometrical theory could be skipped by casual readers, but for the serious-minded individuals, a glossary of terms and a summary of axiom and theorems given without proofs are provided in the Appendix.

The main use of a graph is to place points accurately in a display space so that the coordinates of any point can be accurately read. Any additional utilities such as the reading of distances between points and the gradients of line segments cannot be extended beyond two dimensions using a Cartesian system of coordinates. A good test to determine whether a graph can truely "coordinate" points is by drawing a point which is removed from the main axes and the graphical origin, and to see if the coordinates of this point can be read without ambiguity. You will find that it is impossible to do this using graphs derived from conventional projections such as the isometric, parallel, and perspective projections. Any graph which has a single graphical origin and claims to represent objects beyond two dimensions will fail this test. Bijection in point geometry has been redefined to mean the one-to-one correspondence between points in the n-dimensional Euclidean space (called E^n-space) and patterns of subpoints in the multidimensional Hgram-space (called $H^{p(m)}$). p is the number of graphical display levels and m is the dimensionality of the primitive subspace used to build the Hgram-space. For m = 2, the primitive subspace is actually the Cartesian 2 space. In this category are the 2-D CRT-screen and the printed page. In fact in the Hgram-graph, both p and m can be independently varied thus making the Hgram-space a very versatile display medium. It can be said that with its development, we have effectively breached the dimensional limitations in conventional display media. Are we entering the age of computer-aided hypervisualization, or "CAH" for short? There is now a growing body of researchers in CAH.[1,9–17]

In mapping from the E^n-space to the $H^{p(m)}$-space, the most important single criterion is bijection. The view advanced in this paper is that letter-series sequences should be regarded as points in E^n-space, and in order to visualize them as point-patterns, one has to map them into the multidimensional Hgram-space. Bijection simply means that every point in E^n-space is represented uniquely and unambiguously in the Hgram-space.

Many readers may not be familiar with point geometry and the Hgram system of coordinates. Shorn of its geometrical complexity, the concept itself is quite simple. The reason why the Cartesian system of coordinates and other derivatives of this system cannot display points beyond two dimensions is because of the use of a single graphical origin as the spatial reference. The other obstacle was actually imposed by Euclid when he defined the point in the *Elements* as having no extent and is thus zero-dimensional. Euclid defined a point which cannot be subdivided into subpoints. The author found it necessary to break with this convention in point geometry. With the removal of the constraint, there is theoretically no upper limit to the number of independent axes which could be set up to create the multidimensional Hgram-space. There is bijection between points in E^n-space and the target Hgram-space.

The Cartesian graph is only point-bijective for the 2:2 mapping. Beyond two dimensions, it becomes non-bijective since the number of graphical origins remains fixed at one. In the Hgram graph, the number of graphical origins increases with increasing dimensionality of space displayed. For example, if a point in E^n-space is given by $[x_1, x_2, \ldots x_{n-1}, x_n]$, then this point is

meaning of the sequence and its phrasing still awaits decoding. We humans like to use real world analogies to speculate on the unknown. Hence, there are advocates who speculate that a DNA sequence may be written like a music-score, and others think it is like a text-book or even a computer program. Although DNA sequences are encoded in a linear 1-D string, this does not imply that the information is only 1-D. In fact in abstract geometry, the coordinates of a point in n-dimensional space is expressed as $[x_1, x_2, \ldots x_n]$. If this point is plotted in a generalized Cartesian graph, n independent axes will be needed, and this cannot be visualized in the Cartesian space. As a DNA sequence only needs four types of nucleic acids C, G, A, and T for its bases, one could regard it as equivalent to a large number with a radix of 4. For example Shakespeare's Macbeth could be regarded as an n-dimensional number with the radix of 26 if it is typed only in capitals without using punctuations, numerals and blank spaces. It seems logical to picture a DNA sequence or a protein sequence as a large number which is representable as a point in n-dimensional space. The higher the dimensionality of the display space, the more information can be encoded in a point. Therefore if you map a point from E^n-space to a Cartesian 2 space, there will be a significant loss of information.

Classifications of Vector-Codes

Whenever one uses a vector-code to plot a serial sequence joined head-to-tail, then one has made use of the Hgram graphical format. In a survey of graphical display information, it was surprising to find that the use of the Hgram type of graphical format was more widespread than originally suspected. This means that graphs can be classified purely on their point-bijective property alone. A Cartesian graph is distinct from a Hgram graph because in the former all vectors are referred to one common graphical origin whereas in the latter each vector uses the head of the previous vector as its graphical origin. Table 1 gives the classification of existing and new vector-codes including human languages, music, paintings, and sculptures. For example, in a text-string, each alphabet has its own space and they are joined head-to-tail along a predefined axis which can be horizontal or vertical. Since Gates', Hamori's and Pickover's vector-codes are also plotted in this fashion, they are classified under the Hgram-graph heading. Only displays which convey information are considered for classification. This makes sculpture a borderline case as some sculptures convey more nonsense than sense.

Table 1. Classifications of Graphs.

Hgram Graphs:

Gates' code $H^{p(1*2)}$
Hamori's code $H^{p(2*2)}$
Pickover's code $H^{p(2*2)}$
Huen's code $H^{p(3*4)}$.

Languages: All written human languages, programming languages
and written music, $H^{p(1*2)}$.

Non-Hgram Graphs: (Subset of Hgram Graphs).

Cartesian graphs and derivatives, Isometric, Cavalier, Parallel, Perspective and Engineering projections $H^{1(1*2)}$
paintings $H^{1(1*2)}$
sculptures $H^{1(1*4)}$
Feiner and Besher's n-Vision $H^{p(1*3)}$ [reference 8].

According to point geometry, a 2-D number [x1, x2] can be displayed in one
2-D plane. A 3-D number [x1, x2, x3] will be displayed in a 4-D Hgram graph
using the coordinate representation [x1, x2, x3, 0] but this will require $2 \times$ 2-D
planes. This means that the dimensionality of Hgram graph is always even
so that a 3-D point is equivalent to a 4-D point with the fourth axis unused.
This explains why there is no 3-D subvector in the classifications in Table 1.
Sculpture is classified as $H^{1(1\times4)}$ instead of $H^{1(1\times3)}$.

The Hgram graph $H^{p(m)}$ has two "dimensional parameters" and these are
used to classify other graphs solely using the criterion of bijection for points
from E^n-space to that graphical space. In some vector-codes, such as Hamori's,
Pickover's and Huen's code, the parameter m is not single-valued since more
than one subvector is used to form the vector. For classification, we use the
notation $H^{p(q*r)}$ where q is the number of subvectors used to form the vector
and r is the dimensionality of subpoints used to represent the subvectors, i.e.,
m = q*r. For example Gates' code makes use of one 2-D vector per nucleotide
base in a 2-D Cartesian graph and is classified as $H^{p(1*2)}$ where the magnitude
of p is computed using the axiom in point geometry (see appendix). In the
above example, if a DNA sequence has 100 bases, then n = 100 and p = n/m =
$100/(1*2) = 50$. Then the point in the Hgram-space is a 100-D point and the
space is called $H^{50(1*2)}$. Hamori's code uses two 2-D subvectors to represent
one base and thus q= 2 and r = 2. Therefore Hamori's code is classified as

$H^{p(2*2)}$. Pickover's code is different in that the second vector is represented by color code but is otherwise similar to Hamori's code and is therefore also classified as $H^{p(2*2)}$. All representations are points in multidimensional space even though Gates, Hamori and Pickover referred to them as curves. Under this classification since Huen's code uses three subvectors to represent a protein base and each subvector is 4-D (so that p = 3 and r = 4) therefore the space should be classified as $H^{p(3*4)}$.

As discussed above, the higher the dimensionality of the code, the more information can be encoded in it. Since DNA sequences only require four types of nucleic acids in each base, even Gate's code is adequate, but the curve is highly overlapping and is suitable for identifying global patterns only. Hamori's code is equivalent to Gate's code with a direction bias added which removes the overlapping clusters. However, the H-curve is 3-D requiring at least two views to abstract full information. Pickover's code is also 3-D being based on the vectors of the tetragram and the use of colors for the third direction but the curve is again highly overlapping since no bias is used. Huen's code uses three 4-D subvectors to represent one protein base and thus the point is 12-D. This code incorporates an angular bias in the first subvector to prevent overlapping and all 31 items of information from the IUB table are represented geometrically. Since the representation is that of an n-D point, rotational viewing is unnecessary. The first subvector provides the angular bias to indicate the preponderance of physical/chemical properties and the next two subvectors the distribution of amino acids. If all 31 items of information are plotted in separate display levels, one could even switch on items preferentially to show density patterns of one or more items. This strategy is often used in Hgram graphical displays since subvectors naturally fall into levels of 2-D subspaces.

Design Procedure of Huen's Code for Protein Sequences

The proposed vector-code for protein sequences embodies 31 items of information instead of only four required for DNA sequences. In point geometry, any space beyond two dimensions is a hyperspace. Therefore vector-codes for displays beyond two dimensions should be called hypercodes. Outlined here is the general procedure for the design of Huen's code for mapping of protein sequences into the Hgram graph. The problem is posed as follows:

"According to the IUB standard symbols, there are 26 symbols for amino acids and 5 types of physical/chemical information, making a total of 31 items. Design a vector-code which will map the information bijectively into a two-dimensional target display space. Color-codes may be included as an option but the display must be readable purely from the geometrical attributes. The code is to be tested on the protein sequences of 7 unknown species of seasnakes in order to identify relatedness amongst them purely by visual analysis."

A typical entry for specimen 3 (see Fig. 3) in an ascii-file is given as follows:

MYPAHLLVLLAVCVSLLGASDIPPLPLNLYQFDNMIQCANKGKRATWHY
MDYGCYCGSGGSGTPVDALDRCCKAHDDCYVAEDNGCYPKWTLYSW
QCTENVPTCNSESGCQKSVCACDATAAKCFAEAPYNNKNYNINTSNCQ

Here are the design steps:

Step (1)

To ensure bijective mapping, an Hgram-graph will be used. Decide on the dimensionality of the primitive subspace required to symbolize one protein base. If one 2-D vector is used (such as in Gates' code), then it will be necessary to subdivide the circle into 31 equal parts and this will make it difficult to differentiate between the small angular displacements of adjacent vectors. The dimensionality of Hamori's and Pickover's code are too low to encode 31 items of information. The new code will use 3 × 4-D subvectors where the first subvector is used to differentiate between the physical/chemical properties of the amino acids and the second and third 4-D subvectors used to form distinctive triangles with various base lengths to represent the amino acids. Since there is a maximum of six types of amino acids per physical/chemical grouping, six distinctive triangles are constructed with different base lengths. Right-angled triangles are chosen to bias the figures to the left or the right in order to avoid overlaps. Figure 2(a) shows the geometrical constructions used to build the Huen's code for all 31 items.

To ensure the mapping is readable, one must design the starting and the ending 4-D points as single points even if intermediate points could diverge to form geometrical figures. Figure 2(b) shows the typical linkline diagram of the amino acid serine and the 4-D coordinates of the points, and shows how the geometrical symbol is constructed.

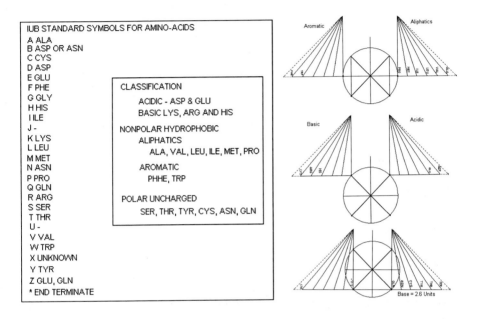

Fig. 2(a). Huen's vector-code for protein sequence.

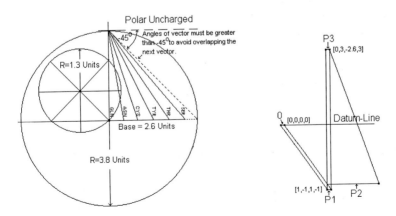

Fig. 2(b). Linkline diagram for Huen's vector-code for serine.

Step (2)

A computer program is written to accept the protein sequence using this Huen's code to convert the protein sequence into a point-pattern. The actual implementation was done on an Intergraph Interact CAD workstation but it can be also implemented in microcomputers with VGA screen format. Experience has shown that the microcomputer version performs faster than the workstation version but the resolution is poorer.

Figure 3 shows the seven species of unknown protein sequences mapped according to Huen's code. The reason why it is classified as $H^{p(3*4)}$ can be seen from the linkline diagram (see Fig. 2(b)) where all three node P1, P2, and P3 are 4-D points. This code needs three nodes per symbol. As far as angular information is concerned, the human eyes can detect easily either 45° or 60° relative angular displacements. In the 4-D code the vectors are formed from three sub-vectors O–P1, P1–P2, and P2–P3. The first subvector O–P1 has adopted 45° increments and there is a choice of six directions which is adequate to symbolize the five types of physical-chemical properties of amino acids. The subvector P1–P2 and P2–P3 together form right-angled triangles of various base-lengths and there are six sizes chosen which can be detected easily by the eyes. Thus altogether 36 possible combinations of information can be encoded in this code for protein sequences even though only 31 symbols are needed. Figure 3 shows the point-patterns of the protein sequences of the toxins of seven species of seasnakes plotted by this vector-code. The curves (or points) are not line-like because they are formed by 4-D points. The inset shows that difference between two nearly identical segments can be easily detected. The non-overlapping appearance is deliberately designed into this code as shown in Fig. 2(a). One must remember that these patterns are points in multidimensional space and not curves. Presumably one could calculate the Euclidean norm between two points which should reflect the relatedness between species. The use of a reduced hyperspace in the form of a multidimensional histogram was reported by Marin[7] using statistical analysis techniques. This method is interesting but the characterization of sequences by the frequencies of occurrence alone means that the mapping is non-bijective.

Physical and chemical properties of the protein sequences in Fig. 3 are easily identified if shown in colors. This greatly helps in the identification of relatedness between species. In monochrome, one will rely on studying the distribution of geometrical shapes which is less easy but still manageable. From Fig. 3, one can discern that species 1 to 3 are closely related. This

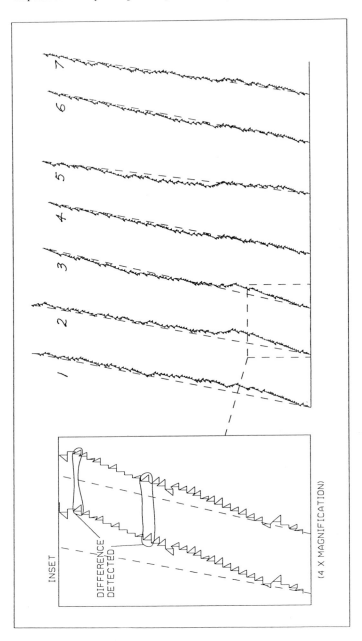

Fig. 3. Protein sequences plotted by Huen's code.

observation can be greatly facilitated by studying the distribution patterns of physical/chemical properties in colors. Not only are they geometrically homologous but also physically/chemically homologous.

No in-depth analyses have been made using the patterns since the current research is in graphical Lyapunov's stability analysis and not biological sequence analysis. The aim is to show researchers that the technique developed in a seemingly unrelated discipline may be adopted for biological sequence analysis.

Clever programming can greatly increase the usefulness of this program. With 26 symbols for amino acids, the use of angular bias to show the preponderance of any one type is not easy. It is far better to plot each type of amino acid in a separate display level so that the display pattern of a particular amino acid can be turned on with the rest switched off for study.

Specifications for the Hgram Software

Generalized packages are not very useful for Hgram graphical applications. This is because the graphical displays are peculiar to the requirements of each discipline which requires customized software. This means that only those who are knowledgeable in a particular discipline will know what features are desirable for program. Several packages of experimental software have been written by postgraduate students and implemented.[9-15] The package used to plot Fig. 3 was written in Pascal and implemented in the Intergraph Interact CAD workstations in our CAD/CAM/CAE centre but as the equipment will soon be replaced by Intergraph Microstations, the package needs to be modified. Another package is currently written in Visual Basic Version 3.0 for Windows and is suitable for stability study rather than genetic sequence visualization. For those who wish to develop their own software, the following specifications are given for Huen's code:

- Equipment: IBM PS/2 or PC-compatibles running on 486 or Pentium microprocessors with VGA or Super-VGA screen format. Colored monitors and mice are mandatory. Alternatively one could use a MacKintosh of equivalent performance.
- Software Packages: Almost any programming language will do including Visual Basic, so long as it has good 2-D graphic support.
- Program Input: The ASCII file will contain either DNA sequence or protein-sequence written in standard IUB symbols.

- Ouput: The screen output can be manipulated using a mouse for zooming, changing of colors, selection of features to be displayed at various assigned levels and also screen printout on a laser printer or a inkjet printer. Even a dot-matrix printer is quite satisfactory.

- Huen's Code: The code contains three 4-D subvectors. The coordinates of each subvector are measured relative to the tip of the previous subvector as the origin. To illustrate this please refer to the linkline diagram shown in Fig. 2(b). For the amino acid SER, the coordinate of point O is [0, 0, 0, 0], point P1 is [1, −1, 1, −1], point P2 is [0, 0, 2.6, 0], and point P3 is [0, 3, −2.6, 3.0]. We could arrange for subvector O–P1 in level 1, P1–P2 in level 2, and P2–P3 in level 3 and assign one color for all three levels to symbolize polar unchargedness. Thus, the subroutine for plotting Huen's code may be expressed as Plotcode (LO, L1, L2, PO, P1, P2, P3) where LO, L1, L2 stands for the three display levels and P1, P2, P3 the coordinates of three 4-D points with respect to point PO which has the initial coordinates [0, 0, 0, 0]. Thus after each plotting the reference point PO should be updated to the new origin given by the algebraic sum of the horizontal and vertical components of the three points P1, P2, and P3. Each time a protein base is read, the coordinates for P1, P2, and P3 can be read from a lookup table according to Huen's code given in Fig. 2(a), or you may design your own geometrical code.

The programming is rather easy since only 2-D graphical plotting is involved even though the points are actually n-dimensional. We have met students who could complete the software in a week and who came without any knowledge of Hgram graph and its theory. In case of difficulty please communicate with the author.

Desirable Properties of Hypercodes

Since high-dimensional space is used to design vector-codes, such codes should be called hypercodes. The desirable properties of hypercodes are listed below:

- Non-overlapping: The masking of information by overlaps can be avoided by design.
- Color option: The code must not rely on colors alone. The codes must be self-sufficient even when printed in black and white without loss of information.

- 2-D graphics: 2-D graphics are preferred to 3-D graphics as the latter takes more CPU time and disk storage. In any case, 3-D graphics are unsuitable for plotting multidimensional Hgram patterns. The use of 2-D graphics is an attractive feature for hypercodes. Programming is very simple even when starting from scratch using 2-D software. Most programmers could create a useable software within a week or less.
- Bijective mapping: The patterns appearing on the computer screen must bear bijective relationship with the letter-series sequences. This property is assured when plotted in the Hgram graphical format.
- No rotational viewing: Since the representation is a point in n-dimensional space, there is no need to abstract information by rotational viewing such as practised in the H-curve by Hamori.[2-4]

Conclusions

This chapter has described a novel way of designing vector-codes using the multidimensional Hgram-graph. A summary of the main points are given below:

- A biological sequence can be viewed as a point in n-dimensional space.
- It is impossible to map information from a high-dimensional space to a low-dimensional space without loss of information using the Cartesian system of coordinates.
- The Hgram system of coordinates ensures the bijective mappings of point information from the Euclidean n-space to the multidimensional Hgram-space.
- The new vector-code for protein sequence is able to embody all the information from the IUB table. If more information is to be embodied, the code has to be redesigned by using an even higher dimensional primitve space.
- The mapping between the protein-sequence and the Hgram pattern is truely in one-to-one correspondence which should be welcome by biologists and geneticists trying to decode these biological sequences.
- The classification gives a unified basis for graphs and should give researchers a good understanding why information is lost in mapping using the Cartesian system of coordinates. Point geometry might provide a geometrical basis for the analysis of biological sequences.

- Point geometry has simplified computer-aided hypervisualization using only 2-D graphics. This effectively has increased the versatility of the computer screen many fold and should be welcome by those interested in visual computing.

Acknowledgements

The author thanks National University of Singapore for the provision of computing facilities and past graduates who have contributed in the research effort. He especially wishes to thank Dr. Tan Tin Wee, presently head of Technet of the National University of Singapore for the provision of the seven protein sequences of local seasnakes which have been plotted in Fig. 3.

References

1. Huen, Y.K., "An introduction to the Hgram (an n-dimensional graph paper)", *Copyright Application Document*, Copyright Registration No. Txu 354026 (USA Library of Congress, 1988).
2. Hamorie, E. and Ruskin, J., "A novel method of representation of nucleotide series especially suited for long DNA sequences", *J. Biol. Chem.* **258**(2), 1318–1327 (1983).
3. Hamorie, E., "Review — Graphic representation of long DNA sequences by the method of H curves — Current results and future aspects", *BioTechniques* **7**(7), (1989).
4. Hamorie, E., Varga, G. and LaGuardia, J.J., "HYLAS: Program for generating H-curves (abstract three-dimensional representations of long DNA sequences)", *CABIOS* **5**(4), 263–269 (1989).
5. Gates, M.A., "A simple way to look at DNA", *J. Theor. Biol.* **119**(3), 319–328 (1986).
6. Pickover, C.A., "DNA and protein tetragrams: Biological sequences as tetrahedral movements", *J. Mol. Graphics* **10**, 1–6 (1991).
7. van Heel, M., "A new family of powerful multivariate statistical sequence analysis techniques", *J. Mol. Biol.* **220**, 877–887 (1991).
8. Feiner, S. and Beshers, C., "Worlds within worlds, metaphors for exploring n-dimensional virtual worlds", *Proc. UIST '90 ACM Symp. on User Interface Software and Technology, Snowbird, UT* (1990) pp. 76–83.
9. Huen, Y.K., "The Hgram graph — its geometrical interpretation and its applications", *Int. J. in Math. Edu. in Sci. and Tech.* **22**(3), 403–418 (1991).
10. Huen, Y.K. *et al.*, "Visualisation of system stability using the Hgram graphical display format", *Proc. of CHEMICA '90, Auckland, New Zealand* **2**, 517–524 (1990).

11. Wawan, S. and Huen, Y.K., "Novel concepts in data displays for computer screens", *Proc. of Int. Conf. on Instrumentation, Measurement and Control* (1991) pp. 177–193.

12. Huen, Y.K., Wawan, S., Loi, K.S. and Allen, R.M., "How will Hgram graphs affect teaching and research methodology in engineering education?" *Proc. of the 3rd Triennial Conf. of the Assoc. for Eng. Edu. in South-East and the Pacific*, Christchurch, New Zealand (1991) pp. 498–503.

13. Huen, Y.K., Wawan, S., Liang, S.S., Lee, W.L. and Ong, C.H., "The Hgram graphs — its applications in multidimensional visualization on computer screens", *Proc. of Int. Conf. on Inf. Eng. (ICIE'91)* (1991) pp. 847–858.

14. Pok, Y.M. and Huen, Y.K., *Visualisation of Hyperobjects in Hgram-Space by Computers, Modern Geometric Computing for Visualisation* (Springer Verlag, 1992) pp. 141–163.

15. Wawan, S., Huen, Y.K., Rangaiah, G.P. and Sim, K.L., "Graphical optimization using Hgram graphical format", *Proc. of Int. Conf. on Opt. Tech. and Appl. (ICOTA '92)*, Vol. 1 (World Scientific, 1992) pp. 667–671.

16. Wawan, S., "Graphical optimization in chemical engineering processes", M. Eng. Thesis, Department of Chemical Engineering, National University of Singapore (1992).

17. Huen, Y.K. *et al.*, "3-D Visualization of helical structures", *Inter-Faculty Seminar Mathematical Applications in Science and Engineering* (1992) pp. 1–21.

Appendix: PG-Axiom and Theorems

Axiom in Point Geometry (PG-axiom):

In an n:m mapping from the E^n-space to the $(p * m)$D-Hgram-space or $H^{p(m)}$, mapping can only be bijective if the number of graphical reference points p (called suborigins) in the target space is increased according to $p = \text{float}(n/m)$ corrected upward to the nearest positive integer.

Remarks: The Hgram system of coordinates used nested 2-D planes as primitives whereas Feiner and Besher[8] used 3-D space as primitives. According to the above axiom if the dimensionality of the display medium is two, it is impossible to use a primitive space of dimensions greater than two without additional geometrical aids such as the use of a dipstick and a waterline by these two workers. Whilst the concept of nested heirachical dimensions is not new, the above axiom allows the establishment of a self-consistent geometry called point geometry.

Theorems (given without proofs*):

Theorem 1: An n:m mapping is always point bijective in both directions, i.e., $f:E^n \to H^{P(m)}$ and $f^{-1}:H^{P(m)} \to E^n$.

Theorem 2: In an n:m mapping, the vertices of a hyperobject in E^n-space is bijectively represented in the $H^{P(m)}$-space.

Theorem 3: Two hypertrajectories will intersect visually in the $H^{P(m)}$-space if they have identical coordinate values.

Theorem 4: Sublines within any level of the $H^{P(2)}$-space is metrisable by bringing their suborigins together but angular displacements is not metrisable.

Theorem 5: In an n:m mapping, the actual geometrical shape of the object can be interpreted from any one of the permutated axial assignments of independent variables.

Theorem 6: Irrespective of the dimensionality of the Hgram space, the only point which has absolute location is the suborigin of the first level or reference subspace.

Theorem 7: In an n:2 mapping, whether by unique axial assignments or multiple axial assignments, the elements of an n-D point may be expressed in the absolute displacement format as recursive nested functions of x_1 or in the relative dispacement format as the difference between two recursive nested functions of x_1.

Theorem 8: In n:m mapping, the absolute position of a subpoint is non-unique but its relative position to its own suborigin is unique.

Theorem 9: When three collinear points A, B, and C, where C is between A and B, are mapped into the Hgram space, then both the betweenness of C and its segment ratio with respect to AB are preserved.

Theorem 11: In an n:2 mapping, where n is even, the linkline in each level can be resolved into two components in the the the principal X- and the principal Y-direction.

Theorem 12: In an n:2 mapping, a n-D point $P'_n = [(x_1, x_2), \ldots (x_{n-1}, x_n)]$ is referred to its own graphical origin $P'_o = [(0, 0)(x_1, x_2, \ldots (x_{n-3}, x_{n-2})]$. If P'_n is a trajectory, then P'_o is also a trajectory.

*Proofs may be found in reference 17.

Glossary

CAH: Abbreviation for computer-aided hypervisualization.

Hgram: Acronym for Huen's diagram.

IUB: A commission that determined standards in biochemical nomenclature and the symbols for amino acids. See *Biochemistry* **9**, 3471 (1970).

Level: A level contains subspaces in the same layer. For example the second level contains all subspaces of the second layer.

Linkline: The line (hatched attribute) joining two subpoints. It is used to group all the subpoints belonging to a point. Linklines can only join subpoints between adjacent levels of subspaces.

Linkline Diagram: A diagram which shows what each linkline of a point represents.

n-D point: A point in the n-dimensional space plotted in Hgram graphical format. The concept can be extended to lines, surfaces, volumes and hyperobjects.

n-D trajectory: An n-D trajectory is the n-D curve joining several n-D points. It contains n/2 sublines collectively referred to as a trajectory.

Point Geometry (PG): A geometry which is valid for point representation only as all higher geometrical abstractions beyond the reference planes are non-metric. (Note: Lines in higher levels or subspaces is metrisable using Theorem 4).

PG axiom: The axiom on which point geometry is founded.

Point-Pattern: The display of a point as an assemblage of subpoints in the Hgram-space in the form of a unique linkline pattern or diagram.

Subpoint: A 2-D point is called a subpoint. It is the primitive from which a point is built. For example, a 10D-point contains $5 * 2$-D subpoints.

Subspace: A subspace is actually the conventional Cartesian plane. It is the primitive from which the Hgram space is built. In Hgram-graph, one subpoint is drawn in one subspace. The first subspace is called the base-plane or reference plane which contains the abolute reference origin. The next subspace is called the second subspace or second level and so on.

Subline: The line joining two subpoints in the same level (solid attribute).

Index

AIDS   119
Alu sequences   vii
amino acid sequences   6, 61, 129,
   145, 146, 175
amino acid tetragrams   xxxviii
amino acids   15
amount of DNA per cell   vi
atomic models   67
Avery, O.   vi

barogram   99
bijective mapping   172
BIOSCI   xxiv
Brown, M.   viii
Brownian motion   97

Campione-Piccardo, José   96
center of mass   96, 103
chaos   xviii, 115
Chaos game representation   113

Chenault, Kelly D.   6
chromosome   vi
classification, biological   21
cluster analysis   52
clustering   52
codons   2, 90, 119, 121, 135
complement-binding domains
   145, 148
Composite discriminator plots   153
Consensus sequences   28

Dalton, Mark W.   119
Database Repositories   xxi
Databases   43
Databases, gel   45
DeFanti, T.   viii
Dendrogram   52
Diagrammatic Representation   84
DNA   xxxviii
DNA Data Bank of Japan   xxi

DNA Vectorgram   xxxvii
DNA walks   114
DRAW program   121
Duvall, Melvin R.   119

electrophoresis   43
Electrophoresis Databases   43
European Molecular Biology Data
    Library   xxi
evolution   28, 158
exons   114
expression-profile   49

faces   xxxix
Federal Research Organizations   xx
Folding of mRNA   121
folding structure   158
Fourier filtering   98
Fourier transforms   97, 98
fractal dimensions   111, 114
fractals   xviii, 96, 102, 112
fractional Brownian motion   97, 98
fruit fly   vii

G-protein-coupled receptors
    145, 148
Gates, M.A.   165, 166
gel electrophoresis   43
GenBank   xxi
Gene Music   72
genetic code   1, 15
Genetic homology   21
Genome Data Base   xxi
Gray code   2

H-curves   97
Hackett, Perry B.   119
Hamori, E.   165, 166

Hayashi, Kenshi   72
helical nets   129
Helical representation   138
helices   16
Hgram system   165, 166
histogram   54
homeobox   vii
Huen's code   171
human genome project   vii
human immunodeficiency virus
    120
human papillomavirus   106
Hurst exponent   98, 111
Hydrophobic Cluster Analysis
    Method   129

image analysis   44
information mass volume   158
Internet   xxxii
Intron-exon junctions   33
introns   114
isosurface   xxxvi

Johnson, Darrin P.   119

"Kitty" representation   12
kringle domains   145, 148
Kuznetsov, D.A.   61

Lévy walks   114, 115
Lim, H.A.   61
Linear- or β-sheet-like
    representation   138
logo, sequence   23
long-range correlations   114
Lyapunov's stability analysis   176

McCormick, B.   viii

Melcher, Ulrich   6
MIDI instruments   75
Miescher, F.   vi
Mizraji, Eduardo   33
Mosaics   49
mRNA   158, 161
mRNA sequences   119
multifractal spectrum   114
Munakata, Nobuo   72
Music   xvii, 72

Newsgroup   xxiv
Ninio, Jacques   33
Noise   97
Nucleic Acid Sequences   33

oncogenes   163
Open Reading Frames   119, 120

PCR Amplification   21
Pickover, Clifford A.   165, 166
Point Geometry Analysis   165
Point Representation   168
polymerase chain reaction   21
polypeptides   43
Power spectrum   xl
Prion   133
protein sequences   6, 12, 61, 129, 145, 175
protein structures   15, 61, 139, 146
Protein Patterns   43
Proteins   51
Puppy representation   6

radius of gyration   103
random   ix
random fractals   97
random nucleotide sequences   158

randomness   34
retrovirus   119
ribosomal RNA   21
Rmaps   49
RNA Folding   158
RNA secondary structure   121
RNA structure   119
RNA tumor viruses   119
RNA viral sequences   158
RNA virus   161
Rogan, Peter K.   21
rRNA   158
Rspot clustering dendrogram   49

Salvo, Joseph J.   21
sandbox algorithm   114
Schneider, Thomas D.   21
Science-Fiction   xxxiii
scientific visualization   viii, xix
secondary structure   16, 121
Sequence Conservation   21
sequence homology   112
sequence logo   23
sheets   16
snRNA   158
Software   xxx, 61, 140
spectral exponent   111
stem-loop structures   158
Stephens, Michael S.   21
structure, protein   15
Swanson, Rosemarie   1, 15
Swanson, Stanley M.   1

Taxonomic classification   21
Tetragrams   97
tetrahedral lattice   93
tetrahedron   85, 96, 99

three-dimensional structures   15
triplet code   1
tRNA   1, 158

vector-codes   165, 166, 169
Vectorgrams   97
vectorial   38
Vectorial coding   38
vectorial representation   33

VisiCoor   61

Williams, Ann   6

Yamamoto, Kenji   158
Yoshikura, Hiroshi   158

Zhang, C.-T.   84
zinc fingers   145, 148